INTRODUCTION TO SEMICONDUCTOR DEVICE MODELLING

INTRODUCTION TO SEMICONDUCTOR DEVICE MODELLING

Christopher M Snowden

Published by

World Scientific Publishing Co. Pte. Ltd.

P O Box 128, Farrer Road, Singapore 912805

USA office: Suite 1B, 1060 Main Street, River Edge, NJ 07661

UK office: 57 Shelton Street, Covent Garden, London WC2H 9HE

British Library Cataloguing-in-Publication Data
A catalogue record for this book is available from the British Library.

First published 1986
First reprint 1998

INTRODUCTION TO SEMICONDUCTOR DEVICE MODELLING

ISBN 9971-50-142-2
ISBN 981-02-3693-X (pbk)

Printed in Singapore by JCS Office Services & Supplies Pte Ltd

To Wendy, Barbara and William

PREFACE

The dramatic developments in the area of semiconductor device technology over the last twenty five years has led to the appearance of many sophisticated solid state components including complex integrated circuits with over half a million active devices on each chip. This in turn has placed increasing demands on the fabrication and design technologies, where it is now necessary to implement sub-micron gate geometries in order to achieve the necessary performance for future devices. The requirement for very high integration levels has also led to the need for very high packing densities, with associated small-scale device dimensions. The exceedingly small size of the active devices means that simplistic ideas of operation are no longer adequate and it has been necessary to develop new techniques and theories to account for the physics and electrical characteristics. Although the area of silicon VLSI (very large scale integration) has been a driving force in recent technology, other types of semiconductor devices have been introduced to meet the requirements for high frequency devices in the communications and radar fields. In particular compound semiconductor devices have been developed to operate in the microwave, millimetre wave and optical frequency ranges. These types of device again frequently have active regions of less than one micron. Clearly, the need to understand the operation of these devices is fundamental to future development and optimal design.

The key to the understanding of semiconductor devices lies in developing satisfactory models to represent the physical and electrical characteristics. Semiconductor devices are modelled using electrical equivalent circuit models and physical device models. This book is dedicated to a study of physical device models derived from the transport physics of semiconductor devices. The importance of physical device modelling techniques will increase as the demand for smaller and more complex devices continues. At the present time equivalent circuit modelling techniques are widely accepted and used throughout industry. Over the next decade it is reasonable to expect that there will be a similar demand for semiconductor device simulations based on the physics of devices. We are already beginning to see the appearance of software packages based on these simulations, which are being taken up by the semiconductor industry.

The state-of-the-art in device simulation is advancing rapidly. In addition to the considerable coverage in the scientific literature, there are now several conferences dedicated to device simulation. New techniques and improved models are continually appearing and the drive towards sub-micron structures has meant that hot electron and quantum effects must now be considered. This array of developments presents the engineer and device physicist with a wide range of tools with which to approach the understanding and design of semiconductor devices.

This book is intended to provide a broad overview of semiconductor device modelling techniques whilst preserving sufficient detail to be of interest to the established device modeller. It contains up-to-date information on techniques, simulations and results. Chapters on the simulation of heterojunction devices and quantum mechanical transport are included to cover recent areas of interest. It is intended that the book should provide a useful introduction to device modelling, presented in an easily absorbed manner, rather than simply acting as a definitive reference. Nevertheless, several algorithms which have been found particularly useful in simulations have been included. The material contained in the book progresses from the introduction of the classical semiconductor equations in Chapter 1 through chapters on analytical and numerical techniques to a chapter on Monte Carlo methods, concluding with Chapter 9 which deals with quantum mechanical theory and its application. Examples of simulation results are included throughout the text to illustrate the various techniques.

I am grateful to my colleagues in the Microwave Solid State Group in the Department of Electrical and Electronic Engineering at the University of Leeds, and in particular Vasil Postoyalko, for many useful discussions. I would also like to express my gratitude to Bruce Hassall for his technical support and advice during the wordprocessing of this text.

Leeds, October 1986 Christopher M Snowden

SYMBOLS

$\mathbf{A}$	vector potential
$\mathbf{B}$	magnetic flux density
B	Bernoulli function
B	electrical susceptance
C	net ionized impurity concentration
C_{ij}	autocorrelation coefficient
C_{OPT}	optical capture-emission rate
C_n	Auger coefficient for electrons
C_p	Auger coefficient for holes
$\mathbf{D}$	electric flux density
D	diffusion coefficient
D_{T_n}	thermal diffusion coefficient for electrons
D_{T_p}	thermal diffusion coefficient for electrons
D_l	diffusion coefficient in drift direction
$\mathbf{E}$	electric field strength (vector)
E	energy (see also ξ)
E	electric field (scalar)
E_c	conduction band energy
E_{fn}	quasi-Fermi energy (electrons)
E_{fp}	quasi-Fermi energy (holes)
E_g	band gap
E_v	valence band energy
$\mathbf{F}$	force (vector)
$\mathbf{F}_E$	external force
$\mathbf{F}_I$	internal force
G	electrical conductance
G	carrier generation/recombination rate
$\mathbf{H}$	magnetic field strength vector
H	thermal generation
H	Hamiltonian
I	total current

J	total electric current density
$\mathbf{J}_n$	electron current density
$\mathbf{J}_p$	hole current density
N_A	concentration of ionized acceptors
N_D	concentration of ionized donors
N_c	effective density of states in conduction band
N_v	effective density of states in valence band
P	position probability density
P	scattering probability
Q	total charge
R	Residual
T	lattice temperature
T_e	electron temperature
V	applied voltage
V	potential energy
V_T	threshold voltage
V_t	thermal voltage (kT/q)
a	incremental distance of a mesh in the x direction
a	a quantity (associated with an average)
α_I	ionization rate
b	incremental distance of a mesh in the y direction
c	speed of light in vacuum ($2.9979\times10^8 ms^{-1}$)
c	specific heat
Δx	incremental distance in x direction
Δy	incremental distance in y direction
ε	permittivity
ε_0	permittivity constant in vacuum ($8.8542\times10^{-12} CV^{-1}m^{-1}$)
ε_r	relative permittivity
f	distribution function
f_T	cut-off frequency
ϕ_n	electron quasi-Fermi potential
ϕ_p	hole quasi-Fermi potential

χ	electron affinity
g_{ds}	drain-source conductance
g_m	transconductance
h	Planck constant ($6.6262 \times 10^{-34} Js^2$)
$\hbar$	reduced Planck constant ($h/2\pi$)
$\mathbf{k}$	momentum vector
k	thermal conductivity
k	Boltzmann constant ($1.3807 \times 10^{-23} J\, K^{-1}$)
m^*	effective mass
μ_n	electron mobility
μ_p	hole mobility
μ_0	permeability constant in vacuum
m_o	electron rest mass ($9.1095 \times 10^{-31} kg$)
n	electron concentration
n_i	intrinsic carrier concentration
n_{ss}	surface state concentration
$\hat{n}$	unit vector (normal)
p	hole concentration
q	elementary charge ($1.6022 \times 10^{-19} C$)
r	random number
$\mathbf{r}$	position vector
ρ	specific mass density
ρ	space charge
ρ	density function
t	time
τ_e	energy relaxation time
τ_m	momentum relaxation time
τ_n	relaxation time of electrons
τ_p	relaxation time of holes
τ_n	electron lifetime
τ_p	hole lifetime
θ	shape function

θ	angle
$\mathbf{v}$	group velocity
$\mathbf{v}$	particle velocity
$\mathbf{v}_n$	electron velocity
$\mathbf{v}_p$	hole velocity
ψ_b	built-in potential
$\mathbf{v}_{nd}$	electron drift velocity
$\mathbf{v}_{pd}$	hole drift velocity
v_{sat}	saturation velocity
v_∞	saturation velocity
$\mathbf{x}$	space vector
x	cartesian co-ordinate
y	cartesian co-ordinate
z	cartesian co-ordinate
ψ	electrostatic potential
ψ	wave function
ψ_b	built-in potential
ξ	average electron energy

CONTENTS

PREFACE vii

SYMBOLS ix

CHAPTER 1 INTRODUCTION

1.1 Modelling 1

1.2 Historical Development of Physical Device Modelling 4

CHAPTER 2 SEMICONDUCTOR CARRIER TRANSPORT EQUATIONS

2.1 The Boltzmann Model 14

2.2 Maxwell's Equations 17

2.3 The Classical Semiconductor Equations 21

2.4 Boundary Conditions 26

2.5 Generation and Recombination 29

2.6 Thermal Conductivity and Heat Flow 33

CHAPTER 3 SOLUTION OF THE SEMICONDUCTOR EQUATIONS CLOSED-FORM ANALYTICAL MODELS

3.1 Solution Techniques for the Semiconductor Equations 37

3.2 Closed-Form Analysis of the Semiconductor Equations 38

3.2.1 Analysis of a pn Junction Diode 38

3.2.2 Analysis of Field Effect Transistor Operation 48

3.2.3 Analysis of MOSFET Operation 54

3.3 Limitations of Closed-Form Analyses 57

CHAPTER 4 NUMERICAL SOLUTION OF THE SEMICONDUCTOR EQUATIONS THE FINITE DIFFERENCE METHOD

4.1 Finite-Difference Schemes 60
4.2 Discretization of the Semiconductor Equations 64
4.3 Methods of Solving Finite-Difference Equations 71
4.4 Boundary Conditions 74
4.5 Examples of Finite-Difference Simulations 77

CHAPTER 5 NUMERICAL SOLUTION OF THE SEMICONDUCTOR EQUATIONS FINITE-ELEMENT METHODS

5.1 The Finite-Element Method and its Application to Semiconductor Device Simulation 101
5.1.1 The Galerkin Method 103
5.1.2 Finite-Element Semiconductor Equations 105
5.2 Examples of Simulations based on the Finite-Element Method 109

CHAPTER 6 SEMICLASSICAL TRANSPORT EQUATIONS HOT ELECTRON EFFECTS

6.1 The Hydrodynamic Semiclassical Semiconductor Equations 119
6.2 Examples of Hot Electron Modelling 127

CHAPTER 7 SIMULATION OF HETEROJUNCTION DEVICES

7.1 Semiconductor Equations for Heterojunctions 137
7.2 High Electron Mobility Transistors 139
7.2.1 Closed-Form Models 141
7.2.2 Numerical Models 143
7.3 Heterojunction Bipolar Transistors 149
7.4 Monte Carlo Simulations 155

CHAPTER 8 THE MONTE CARLO METHOD

8.1	The Monte Carlo Method Applied to Carrier Transport in Semiconductors	160
8.1.1	Equations of Motion, Energy Band Structure and Free Flight	162
8.1.2	Scattering Mechanisms	168
8.2	Treatment of Results	170
8.3	Application of Monte Carlo Simulations	172
8.3.1	Transport Characteristics	172
8.3.2	Application of Monte Carlo Techniques to Device Modelling	179

CHAPTER 9 QUANTUM MECHANICAL EFFECTS AN INTRODUCTION TO QUANTUM TRANSPORT THEORY

9.1	Extension of Semiclassical Transport Concepts to Quantum Structures	190
9.1.1	Quantum Mechanics - Basic Concepts	190
9.1.2	Application of Quantum Mechanics to Semiconductor Device Modelling	194
9.2	Quantum Transport Theory	201
9.2.1	Applications of Quantum Transport Theory	207

APPENDIX 1 NUMERICAL SOLUTION OF THE CURRENT CONTINUITY EQUATION

A.1	Linearised Continuity Scheme	214
A.2	Steady-State Scharfetter-Gummel Schemes	216
A.3	Full Time-Dependent Schemes	218

INDEX 221

CHAPTER 1

INTRODUCTION

Semiconductor devices form the foundation of modern electronics, being used in applications extending from computers to satellite communication systems. A wide variety of devices are available, fabricated from a range of semiconductor materials. The most common active devices found in electronic systems include bipolar and field effect transistors, diodes, thyristors and triacs, although a large number of more specialised types, such as microwave optoelectronic devices are used in many applications. Silicon is the most commonly used semiconductor material for both discrete and integrated devices, although other materials such as gallium arsenide and indium phosphide are becoming more common for specific applications.

1.1. Modelling

In order to characterise a semiconductor device it is necessary to obtain a suitable representation of the electrical and physical processes involved. It is also necessary to develop a description for the processes which cannot be directly observed. This is often achieved by implementing some form of analogy which follows the behaviour of the device as closely as possible within the constraints of the defined operating environment. This process is termed *modelling*.

The process of modelling requires the analysis and/or simulation of the semiconductor device. The term 'analysis' in this context is usually taken to mean the method by which the complex problem of characterising the device is resolved into simpler component parts which allow the required investigation to be achieved in a near exact manner. Simulation may be defined here as the process of imitating the operation of the device by considering the charactersitics of a different analogous system, without resorting to direct practical experimentation on the device. Analysis is often taken to imply the implementation of closed-form mathematical expressions in a model which allows accurate information to be extracted. In contrast, a simulation is assumed to be more approximate in nature (although this need not always be the case), and frequently takes a phenomenological approach.

In all device modelling work it is essential to appreciate the basis and limitations of the model in question. It may appear superfluous to state that it is generally only possible to characterise a process if the effects of the particular phenomena are built in to the model, but it is surprising how often inadequate models are used in practice to represent complex device processes (for example non-stationary effects in very small devices).

Modelling has an important role to play in the design, development and understanding of semiconductor devices. Traditionally the development of solid-state devices has involved a largely empirical design process with many iterations of the fabrication stage being required to achieve the desired specification. Design rules are usually derived from this trial and error approach, which help reduce the number of iterations required for future generations of device. Device modelling techniques can substancially reduce the time and costs required for developing a specific device, by allowing the designer to home in on a suitable geometry and doping profile prior to the fabrication stage. A smaller number of iterations of the fabrication stage are then usually required to achieve the required result. The variability in fabrication parameters (doping profile, mobility and lithography) and present limitations in the accuracy of practical design models (due mainly to computational requirements of the more sophisticated models), mean that at the present time it is unlikely that the development of new devices can be achieved without a limited number of fabrication trials. However, as the control over material quality and repeatability in fabrication techniques improves, along with the rapid developments in computers and device models, and the drive towards smaller and faster devices, device modelling will play an essential role in the development of future devices.

Solid-state device models can be considered in two broad categories - physical device models and equivalent circuit models. Equivalent circuit models, based on the electrical performance of the device, are a popular choice for circuit design applications and for circumstances where rapid evaluation is required. The principal advantage of this technique is that it is generally easy to implement and can be related readily to the electrical performance of the device in question. Equivalent circuit models are for these reasons frequently used in computer aided design packages intended for circuit design and have been used in integrated circuit design methods, where a large number of devices must be simulated simultaneously. However, equivalent circuit modelling techniques are generally limited in their range of application because it is often difficult to accurately relate the model elements to the physical parameters of the device (geometry, doping etc) and because of the bias, frequency dependence and non-linear behaviour of most semiconductor devices. This means that this approach is not suitable for the predicting the charactersitics of new devices and for modelling accurately large-signal operation, such as in oscillators or high speed logic circuits.

In contrast to equivalent circuit models, physical device models are based on the physics of carrier transport, and can provide a greater insight into the detailed operation of the device. Physical device models are not limited by the operating conditions and have been used successfully to analyse dc, transient, large-signal and high frequency operation. They may be used to predict the characteristics of new devices within the constraints of information available on the semiconductor material properties. The major limitation of this technique is that at the present time detailed physical device models usually require substancial amounts of computer time and memory, leading to a lengthy analysis. The analysis of physical device models is usually carried out using numerical techniques implemented on powerful computers. Physical device models have traditionally been within the realm of the device physicist but with continuing advances being made in the area of computing they are becoming more and more attractive to the the device designer. The drive towards smaller scale devices to achieve faster operating speeds has led to increasingly more complex device structures, where the operating characteristics depart those predicted by classical models. Physical modelling allows these devices and other new structures to be rigourously characterised before fabrication.

Physical device models are solved using either bulk carrier transport equations, Boltzmann transport models or quantum transport concepts. Traditionally, a set of phenomenological transport equations based on the first two moments of the Boltzmann transport equation, which assume equilibrium transport conditions, have satisfied most modelling requirements. Rigourous Boltzmann and quantum models have generally been restricted to providing detailed insight into carrier transport physics. The trend towards sub-micron geometry devices, means that it is essential to consider non-equilibrium transport conditions and develop new models. This has led to increased interest in Boltzmann solutions and quantum transport models. Furthermore, the introduction of low-dimensional structures has in turn led to interest in characterising the surface and contact properties of these new devices.

The aim of this text is to introduce the reader to current techniques used in the physical modelling of semiconductor devices. Process modelling and traditional equivalent circuit modelling techniques are covered in detail in many other texts.

1.2. Historical Development of Physical Device Modelling

Prior to the widespread availability of powerful digital computers, solid-state devices were theoretically characterised using closed-form analytical techniques based on approximate solutions to the carrier transport processes. A well known example of this type of analysis was described by Shockley in his paper on unipolar field effect transistors in 1952 [1]. This approach usually proceeds by dividing the device into regions in which simplified linearised approximations are applied, joined by appropriate boundary conditions [1,2,3]. This method was originally applied to one-dimensional models (Shockley's 'gradual channel' model), but was later extended to include two-dimensional effects in both silicon and gallium arsenide devices [4,5,6]. Important effects such as carrier velocity saturation, absent from some of the very early models, were included in later analyses [7,8]. The closed-form analysis technique proved very effective in characterising large geometry unipolar devices and has continued to be used in many applications which take advantage of the relative simplicity and ease of programming inherent in this approach. However, although this approach allows rapid analysis and provides a basic insight into the device physics, it is unsuitable for modelling devices where the transport process is other than largely one-dimensional and where the electric field varies rapidly throughout the device. This means that a closed-form method is unsuitable for modelling sub-micron devices and many planar devices, such as field-effect transistors (FET's) found in a wide range of discrete and integrated forms.

Interest in the numerical simulation of semiconductor devices, using physical device models, began over twenty years ago. In 1964 Gummel successfully demonstrated that this approach could be used to characterise a silicon bipolar transistor, using a one-dimensional steady-state model [9]. The limited computer resources available at this time meant that device simulations had to be restricted to one-dimension. McCumber and Chynoweth demonstrated Gunn instabilities, in what was one of the first reported one-dimensional electron temperature models for a unipolar GaAs sample [10]. De Mari applied one-dimensional numerical models to pn junctions [11,12]. In 1969 Scharfetter and Gummel reported a one-dimensional simulation used to model silicon Read (IMPATT) diodes [13]. Their numerical scheme for accurately solving the continuity equation has now become an established technique in many simulations and is still used in many two-dimensional simulations.

Two-dimensional numerical simulations were developed to obtain a more realistic representation of planar and three terminal devices (one-dimensional models are still used to model devices with a predominantly 'vertical' structure such as is found in pn junction diodes, Schottky barrier varactor diodes and vertical transferred electron devices). Two-dimensional models also allow other important phenomena such as current crowding and high level injection in bipolar junction transistors (BJT's) and short and thin channels in FET's to be investigated, which is not possible for one-dimensional models. Kennedy and O'Brien reported a two-dimensional simulation for silicon junction field-effect transistors (JFET's) in 1970 [14]. A two-dimensional bipolar junction transistor simulation was described by Slotboom in 1973 [15]. Two-dimensional numerical device simulations have been used to provide a valuable insight into the operation of thyristors [16].

Considerable effort has been directed at simulating field effect transistors. In particular metal-oxide semiconductor field-effect transistors (MOSFET's) and metal semiconductor field-effect transistors have received much attention. Earlier efforts concentrated on the JFET [14,17,18], whilst a very large number of more recent simulations have been directed at MOSFET's [for example 19,20,21,22,23] and MESFET's [24,25,26,27,28]. In the case of GaAs and InP FET's, earlier simulations modelled devices with doping levels restricted to below $10^{22}m^{-3}$, typical of some of the earlier devices of this type and because it emerged that there were numerical stability problems with simulations of higher doped devices that are not as evident in silicon FET's.

Three-dimensional simulations have been recently developed to account for three-dimensional effects found in small devices with narrow widths and non-uniformities in the active regions. Small geometry VLSI MOSFET's with channel widths of the order of the gate length cannot be accurately modelled using two-dimensional models and three-dimensional simulations have been used to characterise these devices. This type of model has been used to investigate non-uniformities in the channel, fringing field effects, breakdown voltage and threshold voltage variations [29]. Yoshii et al have applied three-dimensional simulations to a range of semiconductor devices and have demonstrated that three-dimensional effects are significant in many modern devices [30]. The use of three-dimensional models is presently restricted by the computing power available, and it can be expected that this type of model will become more widely used as computers continue to develop.

The majority of modern low-dimensional semiconductor devices are subject to regions of high electric fields, carrier gradients and current densities which give rise to non-equilibrium (hot electron) transport conditions. The traditional transport equations, based on field dependent carrier mobility (often known as 'drift-diffusion' models), do not account for the process by which carriers gain energy from the transport conditions (carrier heating), and are not capable of modelling accurately sub-micron VLSI and other high frequency devices. The increasing interest in developing sub-micron and other low-dimensional structured devices, including ultra-short gate length MESFET's and MOSFET's, heterostructure devices such as high electron mobility transistors (HEMT's) and heterojunction bipolar transistors (HBT's), has led to the need to produce models capable of characterising non-stationary transport processes. There have been several approaches developed to deal with this requirement. Classical modelling techniques based on drift-diffusion equations have been extended to include momentum and energy relaxation effects (the 'semi-classical' approach), and provide a relatively easily evaluated model. Monte Carlo methods have been used extensively to characterise transport processes in short samples of semiconductor material and in simplified device structures [for example 31,32]. The carrier transport characteristics obtained from Monte Carlo techniques are frequently used in semi-classical models.

Many simulations based on classical and more recently semiclassical semiconductor equations have reached a high level of development and are available commercially and as public domain software. Some of the more widely available programs include MINIMOS [33], GEMINI [34], PISCES [35], CADDET [36] and HFIELDS [37] for two-dimensional and WATMOS [29] for three-dimensional MOSFET modelling; SEDAN [38], BIPOLE [39] and LUSTRE [40] are one-dimensional bipolar simulations whilst BAMBI [41] allows two-dimensional modelling of bipolar devices. CUPID [42] has been developed specifically to simulate MESFET's. Powerful simulation packages with considerable flexibility have been developed by some laboratories. For example FIELDAY [43], developed by IBM allows two- and three-dimensional simulations of MOS and bipolar devices. CURRY [44], recently developed at Philips, has been applied to MOSFET simulations and has bipolar capability. At the present time these latter industrial packages are not generally available. Interest in developing flexible simulations with a choice of solution algorithms has led to the introduction of new generation of software packages. ESCAPADE is an example of this type of package which has its own virtual machine with an application specific instruction set allowing the user to determine the algorithm at run-time using a large library of simulation 'subroutines' [45].

Interest in using Monte Carlo techniques to characterise material properties and device operation has grown steadily since Kurosawa first introduced the method into semiconductor modelling in 1966. Its increasing popularity can be attributed to the continuing search for a detailed understanding of semiconductor properties and because of the steady improvement in computer processing power (both speed and available memory). Monte Carlo techniques have been applied to a wide range of semiconductors, including silicon, gallium arsenide, indium phosphide and aluminium galllium arsenide [46,47,48,49], and to the characterisation of a wide range of devices from FET's to optoelectronic devices [50,51,52,53].

The trend towards very small devices, with dimensions of less than 0.1 microns, has meant that the area of quantum transport theory is finding an important application in the study of semiconductor devices as well as in the more established field of solid-state theory. Quantum transport theory has been used in semiconductor theory to verify the range of validity of Boltzmann transport models. It has not been widely used for device modelling because of the difficulty of implementation. However, quantum transport theory has been more recently used to analyse novel device structures where genuine quantum transport phenomena occur. Quantum transport theory has been used to explain the operation of some ultra-small-scale devices [54,55], although the mathematical complexity and computer power required for this approach is formidable. Quantum transport theory will play an important role in understanding the operation of devices operating in the optical frequency regime, where it has been shown that Boltzmann transport theory is not valid [55] It might be expected that as the dimensions of solid-state devices continue to decrease, quantum modelling techniques will become more significant.

REFERENCES

[1] Shockley, W. "A Unipolar 'Field Effect' Transistor", Proc. IRE, pp.1365-1377, November 1952

[2] Grebene, A.B. and Ghandi, S.K, "General theory for pinched operation of the junction-gate FET", Solid State Electron, No.12, p573, 1969 1969.

[3] Pucel, R.A., Haus, H.A. and Statz, H. "Signal and noise properties of gallium arsenide microwave field-effect transistors", Adv. Electron. Electron Phys., 38, p195, 1975

[4] Lehovec, K. and Zuleeg, R., "Voltage-current characteristics of GaAs JFET's in the hot electron range", Solid State Electron.,Vol. 13, pp.1415-1426, 1970

[5] Shur, M. and Eastman, L. "Current-voltage characteristics, small-signal parameters, and switching times of GaAs FETs", IEEE Trans. Electron Devices, ED-25, No.6, pp.606-611, 1978

[6] Turner, J.A. and Wilson, B.L.H, "Implications of carrier velocity saturation in gallium arsenide field effect transistor", Proc. Gallium Arsenide Inst. Phys. Conf. Ser. No.7, Ed. H Strack, Bristol: Institute of Physics, p195, 1968.

[7] Dacey, G.C. and Ross,I.M., "The field effect transistor",Bell Syst. Tech. J., 34, p1149, 1955

[8] Hauser, J.R., "Characteristics of junction field effect devices with small channel length-to-width ratios", Solid State Electron, Vol.10, p577-587, 1967

[9] Gummel, H.K, "A self-consistent iterative scheme for one-dimensional steady state transistor calculations", IEEE Trans. Electron Devices, ED-20, pp.455-465, 1964

[10] McCumber, D.E. and Chynoweth, A.G., "Theory of negative-conductance amplification and of Gunn instabilities in "two-valley" semiconductors", IEEE Trans. Electron Devices, ED-13, pp.4-21, 1966

[11] De Mari, A, "An accurate numerical steady state one-dimensional solution of the p-n junction", Solid State Electron, Vol.11, pp.33-58, 1968

[12] De Mari, A., "Accurate numerical steady-state and transient on-dimensional solutions of semiconductor devices", Report, California Institute of Technology, Division of Eng. and Appl. Science, October 1967

[13] Scharfetter, D.L and Gummel, H.K., "Large-signal analysis of a silicon Read diode oscillator", IEEE Trans. Electron Devices, ED-16, No.1, pp.64-67, 1969

[14] Kennedy, D.P. and O'Brien, R.R., "Computer-aided two-dimensional analysis of the junction field effect-transistor" , IBM J. Res. Dev., Vol.14, pp.95-116, 1970

[15] Slotboom, J.W., "Computer-aided two-dimensional analysis of bipolar transistors", IEEE Trans. Electron Devices, ED-20, No.8, pp.669-679, 1973

[16] Selberherr, S., Analysis and Simulation of Semiconductor Devices, Springer-Verlag, New York, Vienna, 1984, pp.270-284.

[17] Kennedy, D.P. and O'Brien, R.R., "Two-dimensional analysis of JFET structures containing a low conductivity substrate", Electron. Lett., Vol. 7, No.24, pp.714-717, 1971

[18] Himsworth, B., "A computer aided two-dimensional analysis of gallium arsenide and silicon junction field effect transistors", Int. J. Electronics, Vol.31, No. 4, pp.365-371, 1971

[19] Toyabe, T., Yamaguchi, K., Asai, S. and Mock, M.S."A numerical model of avalanche breakdown in MOSFET's", IEEE Trans. Electron Devices ED-25, pp.825-832, 1978

[20] Yamaguchi, K. and Takahashi, S., "Theoretical characterisation and high speed performance evaluation of GaAs IGFET's", IEEE Trans. Electron. Devices, ED-28, No.5, pp.581-587, 1981.

[21] Yamaguchi, K., "A time-dependent and two-dimensional numerical model for MOSFET device operation", Solid-State Electron., Vol.26, No.9, pp.907-916, 1983

[22] Schutz, A., Selberherr, S. and Potzl, H.W., "Numerical analysis of breakdown phenomena in MOSFET's", Proc. NASECODE II, Dublin, Boole Press, pp.270-274, 1981

[23] Oka, H., Nishiuchi, K., Nakamura, T. and Ishikawa, H., "Two-dimensional numerical analysis of normally-off type buried channel MOSFET's", IEEE Proc. Int Electron Devices Meeting, pp.30-33, 1979.

[24] Reiser, M., "A two-dimensional numerical FET model for DC, AC and large-signal analysis", IEEE Trans. Electron Devices, ED-20, pp.35-44, 1973.

[25] Wada, T. and Frey, J."Physical basis of short-channel MESFET operation", IEEE Trans. Electron Devices, ED-26, pp.467-, 1979

[26] Cook, R.K. and Frey, J., "Two-dimensional numerical simulation of energy transport effects in Si and GaAs MESFET's", IEEE Trans. Electron Devices, ED-29, No.6,pp.970-977, 1982.

[27] Snowden, C.M., Howes, M.J. and Morgan, D.V., "Large-signal modeling of GaAs MESFET operation", IEEE Trans. Electron Devices, pp.1817-1824, 1983

[28] Barnes, J.J., Lomax, R.J. and Haddad, G.I., "Finite-element simulation of GaAs MESFET's with lateral doping profiles and submicron gates", IEEE Trans. Electron Devices, ED-23, No.9, pp.1042-1048, 1976

[29] Husain, A. and Chamberlain, S.G., "Three-dimensional simulation of VLSI MOSFET's: The three-dimensional simulation program WATMOS", IEEE Trans. Electron Devices, ED-29, pp.631-638, 1982

[30] Yoshii, A., Kitazawa, H., Tomizawa, M., Horiguchi, S. and Sudo, T., "A three-dimensional analysis of semiconductor devices", IEEE Trans. Electron Devices, ED-29, 184-189, 1982

[31] Ruch, J., "Electron dynamics in short channel field effect transistors", IEEE Trans. Electron Devices, ED-19, pp.652-659, 1972

[32] Bauhann, P.E., Haddad, G.I., and Masnari, N.A., "Comparison of the hot electron diffusion rates for GaAs and InP", Electronic Letters, Vol.9, No.19, pp.460-461, 1973.

[33] Selberrherr, S., Schutz, A. and Potzl, H.W., "MINIMOS - a two-dimensional MOS transistor analyzer", IEEE Trans. Electron Devices, ED-27, pp.1540-1550, 1980

[34] Greenfield, J.A. and Dutton, R.W., "Nonplanar VLSI device analysis using the solution of Poisson's equation", IEEE Trans. Electron Devices, ED-27, pp.1520-1532, 1980.

[35] Pinto, M.R., Rafferty, C.S. and Dutton, R.W., "PISCES II: Poisson and Continuity Equation Solver", Stanford University, Stanford, CA 94305, September 1984.

[36] Toyabe, T. and Asai, S."Analytical models of threshold voltage and breakdown voltage of short-channel MOSFET's derived from two-dimensional analysis", IEEE Trans. Electron Devices, ED-26, p.453-461, 1979.

[37] Baccarani, G., Guerrieri, R., Ciampolini, P. and Rudan, M., "HFIELDS: a highly flexible 2-D semiconductor device analysis program", Proc. NASECODE IV, Dublin, Boole Press, p1., 1985

[38] D'Avanzo, D.C., "One-dimensional semiconductor device analysis (SEDAN)", Stanford University, Integrated Circuits Laboratory, Report G-201-5, 1979

[39] Denton, T.C., "Validation of BIPOLE", Proc. 2nd International Conf. on Simulation of Semiconductor Devices and Processes, Swansea, Pineridge Press, pp.169-181, 1986

[40] Shedlock, C., "LUSTRE: A fast bipolar process and device simulator using analytic methods", MSc. Thesis, Cornell University, 1984

[41] Franz, A.F., Franz, G.A., Selberherr, S., Ringhofer, C. and Markowich, P., "Finite-Boxes: - A generalisation of the finite-difference method suitable for semiconductor device simulation", IEEE Trans. Electron Devices, ED-30, No.9, pp.1070-1082, 1983.

[42] Wada, T. and Frey, J., "Physical basis of short-channel MESFET operation", IEEE J. Solid St. Circuits, SC-14, 398-412, 1979

[43] Buturla, E.M., Cottrell, P.E., Grossman, B.M., and Salsburg, K.A., "Finite-element analysis of semiconductor devices: The FIELDAY Program", IBM J. Res. Dev, 25, pp.218-231, 1981

[44] Polak, S., Schilders, W. and Driessen, "The CURRY algorithm", Proc. 2nd International Conf. on Simulation of Semiconductor Devices and Processes, Swansea, Pineridge Press, pp.131-146, 1986.

[45] Greenough, C., Hunt, C.J., Mount, R.D. and Fitzsimons, C.J., "ESCAPADE: A flexible software system for device simulation", Proc. 2nd International Conf. on Simulation of Semiconductor Devices and Processes, Swansea, Pineridge Press, pp.639-652, 1986.

[46] Fawcett, W. and Rees, H.D., "Calculation of hot electron diffusion rate for GaAs", Physics Letters, Vol.29A, No.10, pp.578-579, August 1979

[47] Ruch, J.G. and Kino, G.S., "Transport properties of GaAs", Pys. Review, Vol.174, No.3, pp.921-931, 1968

[48] Maloney, T.J. and Frey, J., "Transient and steady-state electron transport properties of GaAs and InP", J.Appl.Phys.,48, pp.781-787, 1977

[49] Littlejohn, M.A., Hauser, J.R. and Glissan, T.H., "Velocity-field characteristics of GaAs with $\Gamma_6^c - L_6^c - X_6^c$ conduction band ordering", J.Appl.Phys., Vol.48, pp.4587-4590, 1977

[50] Moglestue, C., "Computer simulation of a dual gate GaAs field-effect transistor using the Monte Carlo method", IEEE Solid State Electron Devices, 3, pp.133-136, 1979

[51] Moglestue, C., "Monte Carlo particle modelling of small semiconductor devices", Comp.Meth.Appl.Mech.Eng.,30, pp.173-208, 1982

[52] Hockney, R.W., Warriner, R.A. and Reiser, M., "Two-dimensional particle models in semiconductor analysis", Electron. Lett., 10, pp.484-486, 1974

[53] Moglestue, C., "Monte Carlo particle simulation of hole-electron plasma formed in a p-n junction", Electron. Lett., Vol.22, pp.397-398, March 1986

[54] Barker, J.R., in Physics of Non-Linear Transport in Semiconductors, ed. K.Ferry, J.R.Barker, and C.Jacoboni, New York, Plenum, pp.127-151, 1980

[55] Barker, J.R., in Physics of Non-Linear Transport in Semiconductors, ed. K.Ferry, J.R.Barker, and C.Jacoboni, New York, Plenum, pp.589-606, 1980

CHAPTER 2

SEMICONDUCTOR CARRIER TRANSPORT EQUATIONS

Carrier transport can be characterised in terms of either classical or quantum physics, although for most cases, the generalised classical approach described by Boltzmann's transport equation is adequate. Semiconductor equations, derived from the Boltzmann transport equation, are the basis of the majority of current device models, where the dimensions of the device geometry are substancially greater than a de Broglie wavelength ($>0.1\ \mu m$). In the case of much smaller devices it is necessary to consider quantum mechanical effects, which are discussed later in this text.

2.1. The Boltzmann Model

Charge particles can be characterised in terms of their position in space **r** and momentum **k** at time t. The density of particles $n(\mathbf{k},\mathbf{r},t)$ may be described in terms of a distribution function $f(\mathbf{k},\mathbf{r},t)$, which is itself a function of phase and momentum space as well as time.

$$n(\mathbf{r},t) = \int f(\mathbf{k},\mathbf{r},t)\, d\mathbf{k} \qquad (2.1)$$

The derivative of the distribution function with respect to time along a particle trajectory **r**, **k** vanishes in the entire phase space. This is expressed implicitly as the Boltzmann transport equation,

$$\frac{d}{dt} f(\mathbf{k},\mathbf{r},t) = 0 \qquad (2.2)$$

which expands to yield,

$$\frac{df}{dt} = \frac{\partial f}{\partial t} + \frac{\partial f}{\partial \mathbf{k}} \cdot \frac{\partial \mathbf{k}}{\partial t} + \frac{\partial f}{\partial \mathbf{r}} \cdot \frac{\partial \mathbf{r}}{\partial t} = 0 \qquad (2.3)$$

The sum of all the forces **F** acting on the particles can be expressed as

$$\mathbf{F} = \hbar \frac{d\mathbf{k}}{dt} \tag{2.4}$$

where the reduced Planck's constant $\hbar = h/2\pi$. The total force **F** is constituted from forces due to external electromagnetic fields $\mathbf{F}_E$ and internal crystal lattice forces $\mathbf{F}_I$,

$$\mathbf{F} = \mathbf{F}_E + \mathbf{F}_I \tag{2.5}$$

The effect of internal forces acting on the distribution function is evaluated statistically by obtaining an internal collision expression as a function of a scattering probability,

$$\frac{\mathbf{F}_i}{\hbar} \frac{\partial f}{\partial \mathbf{k}} = \int d\mathbf{k}'[f(\mathbf{k}')\mathbf{P}(\mathbf{k}',\mathbf{k}) - f(\mathbf{k})\mathbf{P}(\mathbf{k},\mathbf{k}')] \tag{2.6}$$

The group velocity of the particles may be expressed as,

$$\mathbf{v} = \frac{d\mathbf{r}}{dt} \tag{2.7}$$

An integro-differential expression for the Boltzmann equation is obtained by re-arranging equations (2.4), (2.6) and (2.7) and substituting for the appropriate partial derivatives in equation (2.3),

$$\frac{\partial f}{\partial t} + \mathbf{v}.\frac{\partial f}{\partial \mathbf{r}} + \frac{\mathbf{F}}{\hbar}.\frac{\partial f}{\partial \mathbf{k}} = \int d\mathbf{k}'[f(\mathbf{k}')\mathbf{P}(\mathbf{k}',\mathbf{k}) - f(\mathbf{k})\mathbf{P}(\mathbf{k},\mathbf{k}')] \tag{2.8}$$

where $\mathbf{P}(\mathbf{k},\mathbf{k}')$ is the probability that a carrier will be scattered changing its wave vector from **k** to $\mathbf{k}'$.

A rigourous solution of equation (2.3) or (2.8) is formidable, with no closed-form solutions available. Solutions obtained using iterative procedures are still confined to limited number of special cases. It is more useful at this stage to introduce a series of assumptions to simplify the problem. In order to arrive at a more tractable transport equation it is necessary to assume that the scattering probability is independent of external forces and that all scattering processes are elastic. It is assumed that the duration of collisons is much shorter than the average time of motion of a particle. In practice it is usually assumed that the collisons are instantaneous. It is necessary to assume that the external forces $\mathbf{F}_E$

are constant over a distance comparable with the dimensions of the wave packet describing the motion of a carrier. Particle interaction is assumed to be negligible. If the Lorentz force due to magnetic fields is neglected, the external force $\mathbf{F}_E$ due to the electric field is $q\mathbf{E}$ for positive particles and $-q\mathbf{E}$ for negative particles (electrons). The Boltzmann approximation obtained from these assumptions, based on conventional energy band theory is,

$$\frac{\partial f}{\partial t} + \frac{F_E}{\hbar}.\frac{\partial f}{\partial \mathbf{k}} + \mathbf{v}.\frac{\partial f}{\partial \mathbf{r}} = \left(\frac{\partial f}{\partial t}\right)_{coll.} \tag{2.9}$$

where $\mathbf{v}$ is the group velocity of the carriers. This equation is also described as the Liouville equation of motion for a single particle whose distribution function is $f(k,\mathbf{r},t)$. The collison term is more fully expressed as,

$$\left(\frac{\partial f}{\partial t}\right)_{coll.} = \left(\frac{dn}{dt}\right)_{coll} + \left(\frac{d\mathbf{v}}{dt}\right)_{coll} + \left(\frac{d\xi}{dt}\right)_{coll} \tag{2.10}$$

where n is the carrier density, $\mathbf{v}$ the carrier velocity and ξ the carrier energy. The distribution function is usually assumed to be symmetrical in momentum space. A widely used approximation for the collison term is

$$\left(\frac{\partial f}{\partial t}\right)_{coll.} = -\frac{f - f_o}{\tau} \tag{2.11}$$

where f_o is the spherically symmetric solution and τ is the relaxation time. The particle continuity equations can similarly be derived from the Boltzmann equation to yield the form,

$$\frac{\partial n}{\partial t} + \nabla.(n\mathbf{v}) = \left(\frac{\partial n}{\partial t}\right)_{coll} \tag{2.12}$$

By multiplying equation (2.9) by the group velocity $\mathbf{v}$ and integrating the equation over momentum space, the following momentum conservation equation is obtained for electrons,

$$\frac{\partial \mathbf{v}}{\partial t} + \mathbf{v}.\nabla\mathbf{v} + \frac{q\mathbf{E}}{m^*} + \frac{1}{m^* n}\nabla(nkT_e) = \left(\frac{\partial \mathbf{v}}{\partial t}\right)_{coll} \tag{2.13}$$

where m^* is the effective mass, k Boltzmann's constant, T_e electron temperature and **E** electric field. If the distribution function is assumed to have a displaced Maxwellian distribution and equation (2.9) is multiplied by the average energy ξ and integrated over momentum space the following energy conservation equation is obtained,

$$\frac{\partial \xi}{\partial t} + \mathbf{v}.\nabla \xi + q\mathbf{v}.E + \frac{1}{n}\nabla.(n\mathbf{v}kT_e) + \nabla.\mathbf{s} = \left(\frac{d\xi}{dt}\right)_{coll} \tag{2.14}$$

where **s** is the energy flux. The electron energy ξ is given by

$$\xi = 1/2m^*v^2 + \frac{3}{2}kT_e \tag{2.15}$$

The collision terms are given as

$$\left(\frac{dn}{dt}\right)_{coll} = -\,G(\tau_c) \tag{2.16}$$

$$\left(\frac{d\mathbf{v}}{dt}\right)_{coll} = -\,\frac{\mathbf{v}}{\tau_n(\xi)} \tag{2.17}$$

$$\left(\frac{d\xi}{dt}\right)_{coll} = -\,\frac{\xi - \xi_0}{\tau_e(\xi)} \tag{2.18}$$

where G is generation-recombination rate which is a function of the carrier lifetime τ_c. τ_n is the momentum relaxation time and τ_e is the energy relaxation time, both of which are functions of average electron energy. Equation (2.12), (2.13) and (2.14) are obtained by taking the momentum moments of the Boltzmann transport equation. In unipolar simulations the generation-recombination rate G is usually taken as zero.

2.2. Maxwell's Equations

Maxwell's equations can also be manipulated to extract the current continuity equations and an appropriate expression for the Poisson equation. The carrier transport model must satisfy Maxwell's equations:

$$\nabla\times\mathbf{H} = \mathbf{J} + \frac{\partial \mathbf{D}}{\partial t} \tag{2.19}$$

$$\nabla\times\mathbf{E} = -\,\frac{\partial \mathbf{B}}{\partial t} \tag{2.20}$$

$$\nabla . \mathbf{D} = \rho \tag{2.21}$$

$$\nabla . \mathbf{B} = 0 \tag{2.22}$$

where **H** and **E** are the magnetic and electric field strengths respectively and **D** and **B** are the electric and magnetic flux densities respectively. **J** is the total conduction current density and ρ is the electric charge.

The continuity equations may be derived from equations (2.19) and (2.21) by application of the divergence operator,

$$\nabla . (\nabla \times \mathbf{H}) = \nabla . \mathbf{J} + \frac{\partial \rho}{\partial t} = 0 \tag{2.23}$$

(recall that $\nabla . (\nabla \times \mathbf{a}) = 0$). The electric charge ρ may be defined in terms of the positive charge due to hole density p, negative charge due to the electron density n and the charge due to charge defects Q_d,

$$\rho = q(p - n + Q_d) \tag{2.24}$$

where q is the electronic charge. The total conduction current **J** is defined in terms of electron and hole current densities $\mathbf{J}_n$ and $\mathbf{J}_p$ respectively as

$$\mathbf{J} = \mathbf{J}_n + \mathbf{J}_p \tag{2.25}$$

Using equations (2.23), (2.24) and (2.25) two continuity equations for electrons and holes may be conveniently defined as

$$\nabla . \mathbf{J}_n - q . \frac{\partial n}{\partial t} = q . G \tag{2.26}$$

$$\nabla . \mathbf{J}_p + q . \frac{\partial p}{\partial t} = -q . G \tag{2.27}$$

where G is defined as the generation-recombination rate. G is positive for recombination and negative for generation. The continuity equations (2.26) and (2.27) assume that the influence of the charge defects Q_d is negligible ($\partial Q_d / \partial t = 0$). By defining electron and hole current densities $\mathbf{J}_n$ and $\mathbf{J}_p$ in terms of the electron and hole velocities,

$$\mathbf{J}_n = -qn\mathbf{v}_n \tag{2.28}$$

$$\mathbf{J}_p = qn\mathbf{v}_p \tag{2.29}$$

the continuity equations (2.26) and (2.27) are in the same form as those derived from the Boltzmann equation above (equation (2.12)).

Further consideration of equation (2.23) allows us to define the total electric current density $\mathbf{J}_{tot}$ where,

$$\nabla.(\mathbf{J}_{tot}) = \nabla.\left(\mathbf{J} + \frac{\partial \rho}{\partial t}\right) = 0 \tag{2.30}$$

Assuming that the semiconductor has a time-independent permittivity and that polarization due to mechanical forces is negligible the electric flux density can be directly related to the electric field intensity,

$$\mathbf{D} = \varepsilon_o \varepsilon_r.\mathbf{E} \tag{2.31}$$

where epslion sub o is the permittivity of vacuum and epsilon sub r is the relative permittivity of the semiconductor. The permittivity is strictly speaking a tensor quantity, but this is usually neglected in most treatments of the semiconductor equations. Hence using equations (2.21) and (2.31), it is possible to define the total current density as

$$\mathbf{J}_{tot} = \mathbf{J}_n + \mathbf{J}_p + \varepsilon_o \varepsilon_r \frac{\partial \mathbf{E}}{\partial t} \tag{2.32}$$

where the last term on the right-hand side of the equation is known as the displacement current.

The electric field $\mathbf{E}$ is readily related to the electric potential ψ and charge ρ. In order to preserve generality it is useful at this stage to consider fully time variant equations. In order to facilitate the derivation of relationships between $\mathbf{E}$, ψ and ρ, it is useful to introduce the magnetic vector potential $\mathbf{A}$. Since $\mathbf{B}$ has zero divergence (equation (2.22)), it may be derived from the curl of a the vector potential $\mathbf{A}$,

$$\mathbf{B} = \nabla \times \mathbf{A} \tag{2.33}$$

where

$$\nabla.\mathbf{A} = -\frac{1}{c^2}\frac{\partial \psi}{\partial t} \tag{2.34}$$

and

$$\nabla.\mathbf{A} = -\mu\varepsilon\frac{\partial\psi}{\partial t} = -\frac{1}{v^2}\frac{\partial\psi}{\partial t} \tag{2.35}$$

this is known as the Lorentz condition, where v is the velocity of propagation and μ is the permeability here. In free space the velocity of propagation is equal to the velocity of light. In practice for devices operating at microwave frequencies or lower the right-hand side of equation (2.35) is negligible and the equation may be re-written as,

$$\nabla.\mathbf{A} \simeq 0 \tag{2.36}$$

Substituting equations (2.33) and (2.36) into the Maxwell equation (17) yields,

$$\nabla\times\mathbf{E} = -\nabla\times\left(\frac{\partial\mathbf{A}}{\partial t}\right) \tag{2.37}$$

or

$$\nabla\times\left(\mathbf{E} + \frac{\partial\mathbf{A}}{\partial t}\right) = 0 \tag{2.38}$$

The curl of the gradient of a scalar function ψ is identically zero, hence the integral of equation (2.38) yields,

$$\mathbf{E} + \frac{\partial\mathbf{A}}{\partial t} = -\nabla\psi \tag{2.39}$$

Substituting for **E** in equation (2.31) yields,

$$\mathbf{D} = -\varepsilon\frac{\partial\mathbf{A}}{\partial t} - \varepsilon\nabla\psi \tag{2.40}$$

Further substitution of this expression for **D** in equation (2.21) yields

$$\nabla.\left(\frac{\partial\mathbf{A}}{\partial t} + \nabla\psi\right) = -\frac{\rho}{\varepsilon} \tag{2.41}$$

In the case where the permittivity is homogeneous this equation reduces to the Poisson equation

$$\nabla.\nabla\psi = -\frac{\rho}{\varepsilon} = \frac{q}{\varepsilon}(n - p - Q_d) \tag{2.42}$$

and the electric field is expressed as

$$\mathbf{E} = -\nabla\psi \tag{2.43}$$

The Poisson equation and electric field dependence on ψ could have been more simply derived for the case of static fields (the electrostatic case), where $\nabla\times\mathbf{E} = 0$ using $\nabla\times\nabla\psi = 0$, however it is useful to appreciate the effects time dependence on the relationship between $\mathbf{E}$ and ψ. A rigourous solution of the time-dependent equations leads to the well known Helmholtz equation,

$$\nabla^2\psi + \omega^2\varepsilon\mu\psi = -\frac{\rho}{\varepsilon} \tag{2.44}$$

However, the Poisson equation is generally used in semiconductor device modelling.

2.3. The Classical Semiconductor Equations

The classical semiconductor equations, which form the basis of the majority of physical device models, may be obtained from an approximate solution for the first two moments of the Boltzmann Transport Equation. The momentum conservation equation (2.13) is further simplified in the drift-diffusion approximation by assuming that the electron temperature gradient ∇T_e is negligible and that the term $\mathbf{v}.\nabla\mathbf{v}$ is small compared with other terms in the equation. Assuming that there is at least an order of magnitude difference between the device and circuit responses, implies that a quasi-steady-state model is adequate for most purposes and hence it is convenient to assume that $\frac{\partial\mathbf{v}}{\partial t} = 0$. Finally it is assumed that the electron temperature is equal to the lattice temeperature T. Under these circumstances, equation (2.13) for electrons reduces to the more familiar form

$$\mathbf{v}_n = -\mu_n\mathbf{E} - \frac{D_n}{n}\nabla n \tag{2.45}$$

which treats electrons as negatively charged particles of charge $-q$ Coulombs and where the electron mobility μ_n is defined as

$$\mu_n = \frac{q\tau_p}{m^*} \tag{2.46}$$

and the diffusion coefficient for electrons is defined as

$$D_n = \frac{kT\mu_n}{q} \tag{2.47}$$

where k is Boltzmann's constant and T is the lattice temperature. The lattice temperature is generally assumed to be constant in classical models. This representation of carrier transport is often referred to as the 'drift-diffusion' model. The exact form of the diffusion term in the velocity and current density equations has been the subject of considerable debate. The diffusion coefficent D is shown outside the derivative term in these equations, which is the usual form. However, within the context of the Boltzmann model it is often argued that the diffusion coefficient D should be included inside the gradient operator, ∇Dn. The derivation of both of these forms assumes that the distribution function is only slightly disturbed from equilibrium. Highly non-equilibrium conditions are often analysed using a displaced Maxwellian method, where it is difficult to isolate a diffusion term as such. The derivation of velocity equations and the form of the usefulness of diffusion coefficients is further discussed in Chapter 6.

In general terms, the classical carrier transport models require the self-consistent solution of the following equations.

Continuity equations

$$\frac{\partial n}{\partial t} = \frac{1}{q} \nabla . \mathbf{J}_n + G \quad \textit{for electrons} \tag{2.48}$$

$$\frac{\partial p}{\partial t} = -\frac{1}{q} \nabla . \mathbf{J}_p - G \quad \textit{for holes} \tag{2.49}$$

where $\mathbf{J}_n$ and $\mathbf{J}_p$ are the electron and hole current densities respectively and G is the generation-recombination rate.

Current densities

$$\mathbf{J}_n = qn\mu_n\mathbf{E} + qD_n\nabla n \quad \textit{for electrons} \tag{2.50}$$

$$\mathbf{J}_p = qn\mu_p\mathbf{E} - qD_p\nabla p \quad \textit{for holes} \tag{2.51}$$

where μ_n and μ_p are the electron and hole mobilities, D_n and D_p are the carrier diffusion coefficents.

Equation (2.50) and (2.51) are often referred to as the semiconductor mobility equations. The electric field $\mathbf{E}$ is given by

$$\mathbf{E} = -\nabla\psi \tag{2.52}$$

The potential distrubution and electric field are related to the charge in the device by Poisson's equation

$$-\nabla^2\psi = \frac{q}{\varepsilon_o\varepsilon_r}\left(N_D - n + p - N_A\right) \tag{2.53}$$

where ψ is the electrostatic potential, q electronic charge, $\varepsilon_o\varepsilon_r$ permittivity, N_D donor doping density, N_A acceptor doping density, n electron concentration and p hole concentration.

The diffusion coefficients D_n and D_p are usually defined using the well known Einstein relationships,

$$D_n = \frac{\mu_n kT}{q} \tag{2.54}$$

$$D_p = \frac{\mu_p kT}{q} \tag{2.55}$$

where k is the Boltzmann constant and T the lattice temperature. A common variation on these equations is to replace the carrier mobility-electric field products by drift velocity terms (due to the electric field),

$$\mathbf{v}_{nd} = -\mu_n\mathbf{E} \tag{2.56}$$

$$\mathbf{v}_{pd} = \mu_p\mathbf{E} \tag{2.57}$$

where v_{nd} and v_{pd} are the electron and hole drift velocities respectively. Note that the electrons are treated as negatively charged particles (with charge $-q$ Coulombs). This velocity term should not be confused with the total velocities v_n and v_p due to the field and diffusion components defined in equations (2.28), (2.29) and (2.45).

Majority carrier devices such as MESFETs, Schottky varactor diodes and GaAs transferred electron devices (TEDS) are usually analysed using a single species set of the basic transport equations.

The semiconductor equations are often normalised to simplify the expressions, leading to a more straightforward algorithm saving on computer time. The normalised semiconductor equations are usually expressed in the form (see for example [1,2]),

$$\frac{\partial n}{\partial t} = \nabla.\mathbf{J}_n + G \tag{2.58}$$

$$\frac{\partial p}{\partial t} = -\nabla\mathbf{J}_p - G \tag{2.59}$$

$$\mathbf{J}_n = \gamma_n^{-1}(-n\nabla\psi + \nabla n) \tag{2.60}$$

$$\mathbf{J}_p = \gamma_p^{-1}(-p\nabla\psi - \nabla p) \tag{2.61}$$

$$\nabla^2\psi = n - p - N \tag{2.62}$$

where γ_n^{-1} and γ_p^{-1} are the normalised carrier mobilities. The expressions for current densities $\mathbf{J}_n$ and $\mathbf{J}_p$ assume the Einstein relationship for the diffusion coefficients. The normalised semiconductor equations cannot be used in simulations which adopt diffusion-field characteristics other than the Einstein relationship (such as in [3]).

Many simulations use algorithms which are based on the electron and hole quasi-Fermi levels ψ_n and ψ_p rather than directly calculating the electron and hole concentrations,

$$\psi_n = \psi - ln(n) \tag{2.63}$$

$$\psi_p = \psi + ln(p) \tag{2.64}$$

The carrier concentrations n and p in either the original set of semiconductor equations (2.48) to (2.53) or in the normalised equations (2.58) to (2.62) are replaced by these expressions. The quasi-Fermi level approach has been used in both finite-difference and in finite-element numerical solution methods [4,5,6,7]. Quasi-Fermi level models are often used for simulating devices where both types of carrier are involved in the transport process, and in semiconductor devices which have a complex energy band structures (bipolar and heteojunction devices). In steady-state analyses the equations (2.58), (2.59) and (2.62) may be reduced to three coupled elliptic partial differential equations by introducing the new variables ϕ_n and ϕ_p, where

$$\phi_n = \exp\left(-\psi_n\right) \tag{2.65}$$

$$\phi_p = \exp\left(\psi_p\right) \tag{2.66}$$

The resulting set of transport equations is generally easier to solve than those containing the time-dependent continuity equations which are non-linear parabolic partial differential equations.

Boltzmann's statistics have been used to relate the carrier concentrations n and p to the potential ψ, for nondegenerate semiconductors with parabolic band structures thermal equilibrium,

$$n = n_i \exp\left(\frac{q\psi}{kT}\right) \tag{2.67}$$

$$p = n_i \exp\left(-\frac{q\psi}{kT}\right) \tag{2.68}$$

where n_i is the intrinsic carrier concentration at thermal equilibrium and is given by

$$n_i = \sqrt{(np)} = \sqrt{(N_c N_v)} \exp\left(-\frac{E_g}{2kT}\right) \tag{2.69}$$

where N_c and N_v are the effective densities of states in the conduction and valence bands respectively. E_g is the energy band gap of the semiconductor. A non-linear Poisson equation may be derived by substituting for n and p in equation (2.53),

$$-\nabla^2\psi = \frac{q}{\varepsilon_o \varepsilon_r}\left[-n_i \exp\left(\frac{q\psi}{kT}\right) + n_i \exp\left(-\frac{q\psi}{kT}\right) + N_D - N_A\right] \tag{2.70}$$

There have been several methods described for solving this non-linear equation. An efficient technique is described in the paper by Mayergoyz [8].

Semiconductor device models described by these equations assume that the carrier velocities are instantaneous functions of the electric field and that the mobility and diffusion coefficients are functions of the electric field alone. In practice the carriers do not respond instantly to changes in the electric field and the mobility and diffusion coefficients of carriers in semiconductors are tensor quantities, dependent on several parameters in addition to the electric field. More detailed simulations often use mobility and diffusion coefficients which are functions of doping density and temperature [9,10]. The derivation of the continuity and current density equations assumes that the semiconductor has a single-valley, parabolic band structure. In the case of compound semiconductors such as GaAs and InP, which have multi-valley band structures, the analysis is more complicated since each valley is described by its own Boltzmann equation. In order to derive continuity and current density equations for these materials it is necessary to solve a coupled set of Boltzmann equations. Blotekjaer has shown that the electron mobility and diffusion coefficients for compound semiconductors can be expressed as functions of the average velocity in the lower valley, rather than the localised electric field [11].

The current I associated with the contact of semiconductor device models is usually obtained by integrating the total electric current density **J** across the a suitable surface surrounding the contact,

$$I = \int_P \mathbf{J} . ds \tag{2.71}$$

The total current **J** includes both the particle currents $\mathbf{J}_n$, $\mathbf{J}_p$ and displacement current,

$$\mathbf{J} = \mathbf{J}_n + \mathbf{J}_p + \varepsilon_o \varepsilon_r \frac{\partial \mathbf{E}}{\partial t} \tag{2.72}$$

2.4. Boundary Conditions

In addition to the semiconductor equations, physical device models require the domain, boundary and initial conditions governing the device to be specified. The domain of the model is defined by the geometry chosen to represent the actual device. This takes the form of a one-, two- or three-dimensional representation, which may be planar or non-planar in the latter two cases. The number of dimensions which are chosen for the model is largely determined by the device dimensions, geometry, contact and surface properties, and uniformity of field and carrier distributions. Simple device structures, such as those found in vertical two-terminal devices (diodes), with large cross-sectional areas relative to there thickness, may be adequately represented using one-dimensional models, Figure 2.1(a). In contrast, the majority of surface-orientated devices, including those found in integrated circuits, require at least two-dimensional models to account for the multi-dimensional nature of the electric field and carrier distributions Figure 2.1(b). In devices where the width of the active region is much greater than the length, which is usually the case for analogue devices, a two-dimensional model is adequate. However, in some circumstances where the where device width is of the same order as the length, such as in digital VLSI circuits, a three-dimensional model may be required to account for fringing fields and non-homogeneous current flow in the active channel.

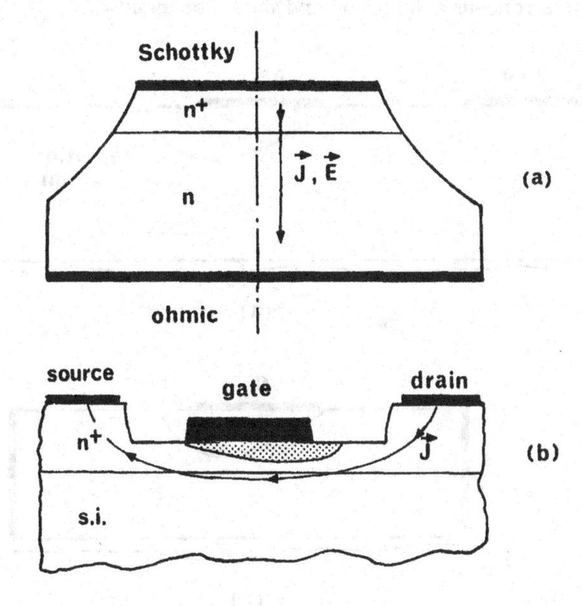

Figure 2.1 (a) vertical device structure
(b) non-planar device structure

The final analysis domain chosen to represent the device is still only an approximation to the real device and the importance of choosing the correct geometry for the model cannot be understated. Surface orientated devices, such as MOSFET's and MESFET's modelled in two-dimensions require planar ohmic contacts on the top surface of the device in order to correctly account for local electric field and space charge effects. However, the model is often simplified to produce a more convenient algorithm and reduce computation time and memory by imposing ohmic boundary conditions on the side-walls at the source and drain perpendicular to the real contacts, Figure 2.2. This can lead to erroneous results in short gate length devices as the field and current density distributions at the ohmic contacts are quite different for the two cases. A further gross simplification, popular in early simulations is the omission of the substrate or buffer layers in these planar devices. This assumes that the conduction current is confined to the active layer and that the substrate plays no part in the operation of the device. This is often an erroneous assumption and should be avoided.

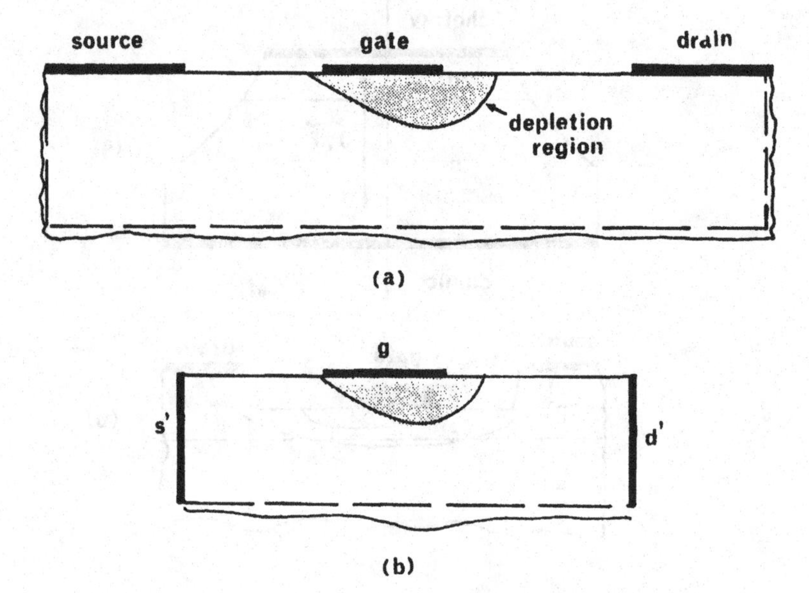

Figure 2.2 Simplified MESFET models showing (a) planar model and (b) frequently used simpliied model which may modify results.

The solution of the partial differential equations which constitute the semoconductor equations requires spatial boundary conditions in ψ, n and p and suitable initial conditions. Surface and contact properties provide the boundary conditions. The presence of current-free boundary regions as well as conducting contacts on the surfaces leads to mixed boundary conditions. Ohmic contacts are usually described by Dirichlet boundary conditions where potential and carrier concentrations are pre-defined at the contacts. Schottky contacts may be modelled using Dirichlet conditions which approximate the reverse bias condition or differential (Neumann) boundary conditions based on thermionic emission theory. The absence of current flow through surfaces may be modelled by assuming that the potential and carrier gradients normal to the surface are zero, if the surrounding media is assumed to have zero relative permittivity. The effects of surface potentials and surface recombination may also be included as boundary conditions. The influence of surface effects on the operation of semiconductor devices can be significant in some circumstances.

The initial conditions required to solve the Poisson and current continuity equations may be satisfied by setting the potential ψ to zero (except at contacts) and the carrier density to the doping density throughout the device. The numerical solution (either time-domain or steady-state), will evolve successfully from these initial conditions, although the solution may take some time to emerge. A considerable time saving can be achieved in successive simulations of the same device by using steady-state distributions obtained from previous solutions as the initial conditions for the next simulation. This technique requires the potential and carrier densities to be stored at the end of the first solution.

2.5. Generation and Recombination

The term G in equations (2.26), (2.27), (2.48) and (2.49) describes the generation-recombination rate. The relative significance of generation-recombination effects depends on the particular device in question. For example, in bipolar devices, such as pn junction diodes and bipolar junction transistors, generation and recombination strongly influence the operation of the device. Unipolar devices, such as MESFET's and Schottky barrier diodes, can be modelled for most operating conditions without accounting for generation-recombination effects. Many MOSFET models also omit generation-recombination rates and obtain satisfactory results. However, it is necessary to incorporate suitable generation-recombination models when investigating high-field and breakdown phenomena, even in unipolar devices.

Consider a homogeneously doped sample of semiconductor material, which is in thermal equilibrium. The concentration of electrons and holes will vary continuously because of generation and recombination processes driven by the thermal energy. In equilibrium, a dynamic balance exists between the generation and recombination rates. This leads to equilibrium electron and hole concentrations n_o and p_o respectively, where

$$n_i^2 = n_o p_o \tag{2.73}$$

and n_i is the intrinsic carrier density. In order to maintain a steady-state distribution in the presence of an external stimulus, the generation and recombination process adjusts to maintain a balance in the numbers of electrons and holes. If an excess of carriers is generated, recombination will dominate, whereas if there is a decrease in the number of carriers, generation will prevail to redress the balance.

Generation-recombination can be considered as result of several distinct processes. The principal mechanisms which cause generation-recombination are thermal (phonon transitions), three-particle interactions (Auger recombination), impact ionization, surface recombination and photon transitions (optical), although other less significant quantum mechanisms also contribute to the overall process. The nett generation-recombination rate can be expressed as the sum of the individual contributions,

$$G \simeq G_{thermal} + G_{Auger} + G_{impact} + G_{surf} + G_{optical} \tag{2.74}$$

The thermal contribution to generation and recombination proceeses, due to phonon transitions occuring as a result of traps, is usually characterised by the Shockley-Read-Hall model. This mechanism is termed an indirect process since it involves a trap centre in the energy band-gap with associated two-stage capture and emission processes. The net generation-recombination rate for the Shockley-Read-Hall model is given by,

$$G_{thermal} = \frac{n_i^2 - pn}{\tau_n(p + p_t) + \tau_p(n + n_t)} \tag{2.75}$$

where τ_n and τ_p are the electron and hole lifetimes respectively. The carrier concentrations n_t and p_t depend on the position and occupancy of the traps. For trap centres in the middle of the band gap,

$$n_t = p_t = n_i \tag{2.76}$$

n_i is commonly substituted for n_t and p_t in equation (2.75). The lifetimes τ_n and τ_p are often defined in terms of the capture rates per carrier,

$$\tau_n = \frac{1}{vN_t a_n} \tag{2.77}$$

$$\tau_p = \frac{1}{vN_t a_p} \tag{2.78}$$

where v is the thermal velocity, N_t the concentration of traps, a_n and a_p the capture cross-sections for electrons and holes respectively. At low doping levels N_t is constant and τ_n and τ_p are independent of doping density. At high doping levels, additional generation-recombination centres occur and the lifetimes are reduced. The lifetimes τ_n and τ_p typically lie in the range 100 ns to 5 μs. A more detailed model for thermal generation-recombination has been developed by Dhariwal, Kothari and Jain [12]. Surface generation-recombination may be modelled using a modified form of equation (2.75), where the carrier lifetimes are replaced by reciprocals of surface recombination velocities (of the order of $10^{-2}\ m^2 s^{-1}$).

Auger recombination involves three particles and involves the recombination of an electron-hole pair and the emission of energy to a third particle. Auger recombination may be considered in terms of direct band-gap generation-recombination (where carriers move directly across the band-gap) and indirect processes involving trap centres. The net Auger generation-recombination rate is given by,

$$G_{Auger} = (n_i^2 - pn)(nC_n + pC_p) \tag{2.79}$$

where C_n and C_p are the Auger coefficients, which are of the order of $3\times10^{-43}\ m^6 s^{-1}$ and $10^{-43}\ m^6 s^{-1}$ respectively at room temperature. Auger recombination becomes significant in regions where there are simultaneously high electron and hole concentrations.

The third contribution to generation-recombination processes is due to impact ionisation. This is a generation process, which is the reverse of Auger recombination. Electron-hole pairs are generated by carriers moving directly across the band-gap. The total generation rate is given by,

$$G_{impact} = \alpha_n \frac{|\mathbf{J}_n|}{q} + \alpha_p \frac{|\mathbf{J}_p|}{q} = \frac{1}{q}(\alpha_n |\mathbf{J}_n| + \alpha_p |\mathbf{J}_p|) \qquad (2.80)$$

where α_n and α_p are the ionization coefficients and $\mathbf{J}_n$ and $\mathbf{J}_p$ are the electron and hole current densities. The ionization coefficients depend on the electric field component E parallel to the direction of current flow,

$$\alpha_n = A_n \exp\left[-\left(\frac{E_{n_{crit}}}{E}\right)^{\beta_n}\right] \qquad (2.81)$$

$$\alpha_p = A_p \exp\left[-\left(\frac{E_{p_{crit}}}{E}\right)^{\beta_p}\right] \qquad (2.82)$$

where the constants A_n and A_p lie in the range $10^7\ m^{-1}$ to $2\times10^9\ m^{-1}$ depending on the material (typically $10^8\ m^{-1}$ for Si). The constants β_n and β_p are usually taken to be unity. The critical fields $E_{n_{crit}}$ and $E_{p_{crit}}$ lie in the range 10^7 to $6\times10^8\ Vm^{-1}$ depending on the material and choice of model (typically $\simeq 2\times10^8\ Vm^{-1}$ for Si and GaAs when $\beta=1$). The ionization rates are zero when the product $\mathbf{E}.\mathbf{J}_{n/p}$ is zero (indicating that there is no field acting in the direction of current flow). Impact ionization rates have been extensively investigated using Monte Carlo techniques and experimental data.

The final mechanism which causes generation-recombination is the process of photon transition. This process involves the direct transition of carriers between the valence and conduction bands. Electrons are excited to the conduction band from the valence band by gaining energy from incident photons. Alternatively electrons lose energy (E_g) which is emitted as a photon and moves from the conduction band to the valence band. This process is significant in narrow band-gap and direct band-gap semiconductors such as GaAs, but has negligible effect in in semiconductors such as silicon. The net generation-recombination rate due to photon transition is given by,

$$G_{optical} = C_{opt}(n_i^2 - pn) \qquad (2.83)$$

where C_{opt} is the optical capture-emission rate.

2.6. Thermal Conductivity and Heat Flow

The electrical charecteristics of semiconductor devices are often a strong function of temperature. The effect of temperature distribution has an effect on all semiconductor devices, but is particularly significant for power devices where increased power dissipation can cause thermal runaway. A complete model of a semiconductor device should account for thermal effects which may play a crucial role in determining the device characteristics.

Thermal effects can be incorporated into device models by solving the heat flow equation,

$$\rho c \frac{\partial T}{\partial t} = \nabla . k(T) . \nabla T + H \tag{2.84}$$

where ρ is the density and c is the specific heat capacity of the semiconductor material. ρ and c are usually assumed to be no temperature dependence. H and $k(T)$ represent the locally generated heat and the thermal conductivity respectively.

Time-dependent simulations require a full solution of equation (2.84). However, for steady-state simulations, where thermal equilibrium is expected (and there is no likelihood of thermal runaway), equation (2.84) may be simplified by assuming that the partial derivative of temperature with respect to time vanishes. This considerably simplifies the solution, but can only be used for conditions of thermal equilibrium.

The presence of temperature gradients through the device leads to induced currents which must be accounted for in the carrier current density equations. The presence of a temperature gradient ∇T requires current density equations of the form,

$$\mathbf{J}_n = qn\mu_n \mathbf{E} + qD_n \nabla n + qnD_{Tn} \nabla T \tag{2.85}$$

$$\mathbf{J}_p = qp\mu_p \mathbf{E} - qD_p \nabla p - qpD_{Tp} \nabla T \tag{2.86}$$

The thermal diffusion coefficients D_{Tn} and D_{Tp} are given by the approximations [13].

$$D_{Tn} \simeq \frac{D_n}{2T} \tag{2.87}$$

$$D_{Tp} \simeq \frac{D_p}{2T} \tag{2.88}$$

The thermal diffusion coefficients are a function of doping density and the expressions given above for intrinsic semiconductors require modifying for heavily doped material (where D_T can increase by upto a factor of five). Exact expressions for the thermal diffusion coefficients are not available at the present time.

The thermal conductivity $k(T)$ follows a non-linear realtionship with temperature. A commonly used relationship for silicon is [14]

$$k(T) = (1.65\times10^{-8}T^2 + 1.56\times10^{-5}T + 0.03)^{-1} \tag{2.89}$$

The coefficients of the quadratic equation are modified for other semiconductors. The thermal conductivity decreases at high doping levels (in contrast to what might be expected).

Several models are available for the thermal generation H. The simplest model accounts for carrier heating using,

$$H = \mathbf{E}.(\mathbf{J}_n + \mathbf{J}_p) \tag{2.90}$$

This model looses its validity in regions where the field-current product is negative where it predicts cooling. A superior model, which also accounts for energy transfer to the lattice through generation and recombination, is based on the expression [15],

$$H = \nabla.\left(\frac{E_c}{q}\mathbf{J}_n + \frac{E_v}{q}\mathbf{J}_p\right) \tag{2.91}$$

where E_c and E_v are the conduction and valence band-edge energies respectively. In the case of non-degenerate semiconductors this equation simplifies to,

$$H = \mathbf{E}.(\mathbf{J}_n + \mathbf{J}_p) + G.E_g \tag{2.92}$$

where G is the generation-recombination rate and E_g is the energy band-gap. The generation and loss of heat due to generation-recombination is accounted for by the last term in equation (2.92), in contrast to the simple and often inappropriate form of equation (2.90).

REFERENCES

[1] De Mari, A. "An accurate numerical steady-state one-dimensional solution of the p-n junction", Solid-State Electron., Vol. 11, pp.33-58, 1968

[2] Slotboom, J.W., "Iterative scheme for 1- and 2-dimensional D.C.-transistor simulation", Electron. Lett., 5, pp.677-678, 1969

[3] Wada, T. and Frey, J., "Physical basis of short-channel MESFET operation", IEEE Solid-State Circuits, Vol.SC-14, No.2, pp.398-412, 1979

[4] De Mari, A., "Accurate numerical steady-state and transient one-dimensional solutions of semiconductor devices", Report, California Institute of Technology, Division of Eng. and Appl. Science, October 1967

[5] Slotboom, J.W., "Computer-aided two-dimensional analysis of bipolar transitors", IEEE Trans. Electron Devices, Vol.ED-20, No.8, pp.669-679, 1973

[6] Sutherland, A.D, "On the use of overrelaxation in conjunction with Gummel's algorithm to speed the convergence in a two-dimensional computer model for MOSFET's", IEEE Trans. Electron Devices, ED-27, pp.1297-1298, 1980

[7] Adachi, T., Yoshii, A. and Sudo, T., "Two-dimensional semiconductor analysis using finite-element method", IEEE Trans. Electron Devices, Vol.ED-26, No.7, pp.1026-1031, 1979

[8] Mayergoyz, I.D., "Solution of the non-linear Poisson equation of semiconductor device theory", J. Appl. Phys., 59, pp.195-199, 1986

[9] Freeman, K.R. and Hobson, G.S., "The V_{fT} relation of CW Gunn effect devices", IEEE Trans. Electron Devices, Vol.ED-19, No.1, pp.62-70, 1972

[10] Snowden, C.M., Howes, M.J. and Morgan, D.V., "Large-signal modeling of GaAs MESFET operation", IEEE Trans. Electron Devices, ED-30, pp.1817-1824, 1983

[11] Blotekjaer, K., "Transport equations for electrons in two-valley semiconductors", IEEE Trans. Electron Devices, Vol.ED-17, No.1, pp.38-47, 1970

[12] Dhariwal, S.R., Kothari, L.S. and Jain, S.C., "On the recombination of electrons and holes at traps with finite relaxation time", Solid-State Electron, Vol.24, No.8, pp.749-752, 1981

[13] Stratton, R., "Semiconductor Current-Flow Equations (Diffusion and Degeneracy)", IEEE Trans. Electron Devices, ED-19, No.12, pp.1288-1292, 1972

[14] Glasbrenner, C.J. and Slack, G.A., "Thermal conductivity of silicon and germanium from 3K to the melting point", Physical Review, Vol.134, No.4A, A1058-A1069, 1964

[15] Adler, M.S., "Accurate calculations of the forward drop and power dissipation in thyristors", IEEE Trans. Electron Devices, ED-25, No.1, pp.16-22, 1978

CHAPTER 3

SOLUTION OF THE SEMICONDUCTOR EQUATIONS
CLOSED-FORM ANALYTICAL MODELS

3.1. Solution Techniques for the Semiconductor Equations

The semiconductor equations consist of a set of partial differential equations which must be solved subject to a pre-defined set of boundary conditions over a specified domain. Although generalised solutions are not available for all devices there are many closed-form analytical solutions available for a wide variety of devices. These closed-form solutions are more suited to lower frequency devices with predominantly one-dimensional field and carrier profiles. They are severely limited in their range of application and accuracy because of the multi-dimensional non-linear nature of most modern devices. A more generalised method of solution frequently applied to the semiconductor equations is to solve them using numerical techniques. The latter approach requires considerably more computer time than the closed-form methods, but usually produces more accurate results and provides greater flexibility.

Numerical techniques are often used to solve the set of non-linear partial differential equations, which constitute the semiconductor equations. However, in many circumstances it is possible to simplify the model and transport equations to an extent which allows useful closed-form expressions to be extracted which describe the electrical behaviour of the device. This approach was used extensively prior to the advent of modern computers because of the difficulty in obtaining numerical solutions. Current semiconductor device technology makes use of very small geometries and it is often inappropriate to apply a closed-form analysis to these structures because of the complex nature of the transport process. Nevertheless, closed-form analytical techniques continue to provide a useful insight into the operation of many devices.

3.2. Closed-Form Analysis of the Semiconductor Equations

The majority of closed-form solutions of the semiconductor equations are based on approximate solutions of the Poisson and current continuity equations. The exact form of the solution depends on the device in question and the degree of sophistication varies from simple analyses applied to junction diodes through to complex models for unipolar fet structures. Closed-form methods are usually restricted to steady-state (dc) analysis with static field, potential and charge distributions.

In order to illustrate the type of approach used in this type of analysis closed-form solutions will be derived for three common device structures - the pn junction diode, unipolar field effect transistor and the MOSFET.

3.2.1. Analysis of a pn Junction Diode

The pn junction diode is defined as a semiconductor device comprising a length of semiconductor material where the net doping $N(x) = (N_A - N_D)$ changes from p-type ($N_A > N_D$) to n-type ($N_A < N_D$). The point where $N_A = N_D$ is defined as the junction. The precise behaviour of the device and its terminal electrical characteristics depends on the exact form of the doping profile. The two most important forms of diode which are considered here are the abrupt junction and linearly graded junction, Figure 3.1.

The case of the abrupt junction will be considered initially. The model assumes that the diode junction is fully depleted of carriers and that the boundaries of the depletion region are abrupt. It also requires that the carrier concentration at the depletion boundaries and contacts are known. The problem may be reduced to one-dimension, yielding a Poisson equation of the form

$$\frac{d^2\psi}{dx^2} = -\frac{q}{\varepsilon}(N_A(x) - p(x) - N_D(x) + n(x)) \tag{3.1}$$

For an abrupt junction, with $x = 0$ at the junction, the following boundary conditions are assumed,

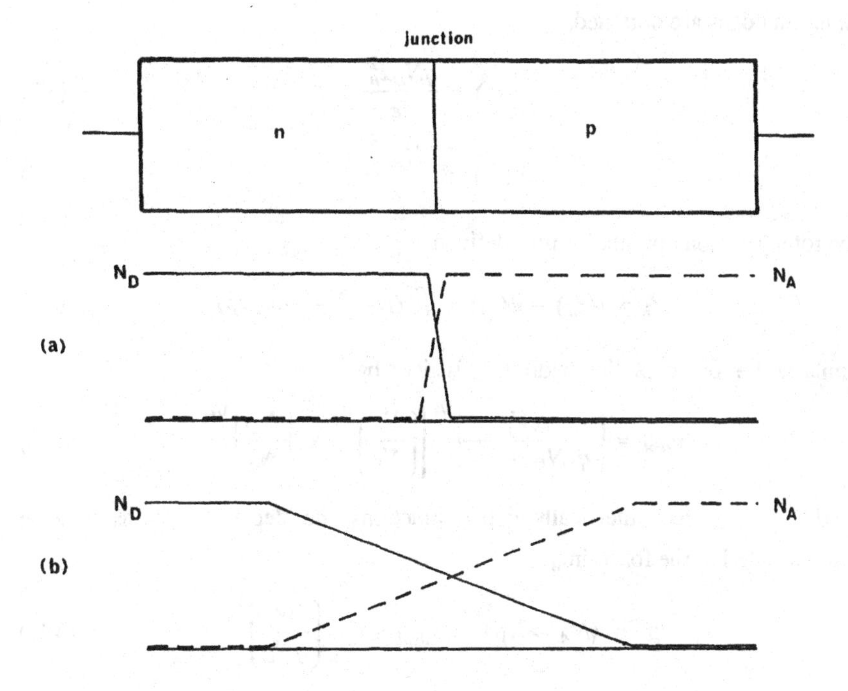

Figure 3.1 (a) abrupt and (b) linearly graded pn junctions

$$\frac{d\psi}{dx} = 0 \quad at\ x = x_n\ (n\ region) \tag{3.2}$$

and

$$\frac{d\psi}{dx} = 0 \quad at\ x = -x_p\ (p\ region) \tag{3.3}$$

hence the electric field E follows the functions

$$E = -\frac{d\psi}{dx} = -\frac{q}{\varepsilon}N_D(x_n - x) \quad \text{for } x > 0 \tag{3.4}$$

$$E = -\frac{d\psi}{dx} = -\frac{q}{\varepsilon}N_A(x_p + x) \quad \text{for } x < 0 \tag{3.5}$$

The maximum value of the electric field is given by

$$E_{\max} = \frac{q}{\varepsilon}N_D x_n = \frac{q}{\varepsilon}N_A x_p \tag{3.6}$$

If it is assumed that $\psi(0) = 0$ then by integrating equation (6), the potentials at the depletion edges are obtained,

$$\psi(x_n) = \frac{qN_D x_n^2}{2\varepsilon} \tag{3.7}$$

$$\psi(x_p) = \frac{qN_A x_p^2}{2\varepsilon} \tag{3.8}$$

The total junction potential is thus defined as

$$V_j = \psi(x_n) - \psi(x_p) = \frac{q}{2\varepsilon}(N_D x_n^2 - +N_A x_p^2) \tag{3.9}$$

Similarly the total depletion width x_{depl} is given by

$$x_{depl} = \left(\frac{2\varepsilon V_j}{q(N_D + N_A)}\right)^{1/2}\left[\left(\frac{N_A}{N_D}\right)^{1/2} + \left(\frac{N_D}{N_A}\right)^{1/2}\right] \tag{3.10}$$

In the case of asymmetrically doped junctions the depletion widths may be approximated to the following,

$$p^+n \;\; (N_A \gg N_D) \qquad x_{depl} \simeq x_n = \left(\frac{2V_j}{qN_D}\right)^{1/2} \tag{3.11}$$

$$n^+p \;\; (N_D \gg N_A) \qquad x_{depl} \simeq x_p = \left(\frac{2V_j}{qN_A}\right)^{1/2} \tag{3.12}$$

where the depletion region extends mainly into the lightly doped semiconductor. The above analysis assumes that the device is in zero bias conditions. If the device is biased with an externally applied voltage V, the variation in depletion widths x_n, x_p and x_{depl} are obtained by replacing V_j by $V_d = V_j - V$ where V is positive for forward bias and negative for reverse bias.

In the case of a linearly graded junction shown in Figure 3.2 the doping density is controlled according to the relationship,

$$(N_D - N_A) = ax \tag{3.13}$$

where a is the doping gradient. The Poisson equation thus becomes,

$$\frac{d^2\psi}{dx^2} = -\frac{q}{\varepsilon}ax \tag{3.14}$$

For the case where the junction is symmetrical $x_n = x_p = \frac{x_{depl}}{2}$. The electric field

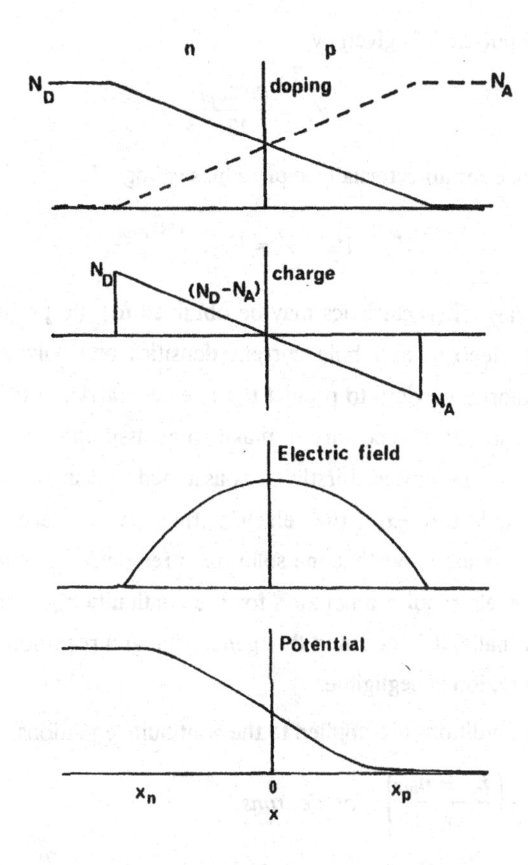

Figure 3.2 Linearly graded junction - charge, electric field and potential distributions

E is obtained by integrating this equation to yield,

$$E = -\frac{\partial\psi}{\partial x} = \frac{qax_{depl}^2}{8\varepsilon} \quad at\, x = 0 \tag{3.15}$$

The potential function $\psi(x)$ is obtained by further integration to yield,

$$\psi = \frac{qa}{2\varepsilon}\left[\left(\frac{x_{depl}}{2}\right)^2 x - \frac{x^3}{3}\right] \tag{3.16}$$

Hence

$$\psi(x_n) = -\psi(x_p) = \frac{qax_{depl}^3}{24\varepsilon} \tag{3.17}$$

and the junction potential is given by

$$V_j = \frac{qax_{depl}^3}{12\varepsilon} \tag{3.18}$$

In the general case for an externally applied bias voltage V,

$$V_d = V_j - V = V_j = \frac{qax_{depl}^3}{12\varepsilon} \tag{3.19}$$

The current voltage characteristics may be obtained for the pn junction diode by introducing the electron and hole current densities and solving the continuity equation for minority carriers to predict the injected carrier profiles. In order to simplify the solution, it is necessary to make some assumptions which govern the carrier transport in the device. Firstly, it is assumed that in regions of the device where $x > x_n$ and $x < -x_p$, the electric field E is zero. Following this approximation, it is assumed that the solution is restricted to low injection levels (high injection levels require a solution for the continuity equations in all regions of the device). Finally, it is assumed that generation and recombination of carriers in the depletion region is negligible.

If the following conditions are applied to the continuity equations,

$$G = -\left(\frac{n_p - n_{po}}{\tau_n}\right) \quad \textit{for electrons} \tag{3.20}$$

$$G = -\left(\frac{p_n - p_{no}}{\tau_p}\right) \quad \textit{for holes} \tag{3.21}$$

and

$$\frac{\partial n_p}{\partial t} = 0 \qquad \frac{\partial p_n}{\partial t} = 0 \tag{3.22}$$

the continuity equations may be simplified to

$$D_n \frac{d^2 n_p}{dx^2} = \frac{n_p - n_{po}}{\tau_n} \tag{3.23}$$

$$D_p \frac{d^2 p_n}{dx^2} = \frac{p_n - p_{no}}{\tau_p} \tag{3.24}$$

The current density equations for carriers at the edges of the depletion region are given by

$$J_n = qD_n\left(\frac{dn_p}{dx}\right)_{x=x_n} \tag{3.25}$$

$$J_p = qD_p\left(\frac{dp_n}{dx}\right)_{x=x_p} \tag{3.26}$$

If it is assumed that the minority carrier concentrations at $x = x_n$ and $x = x_p$ are defined by the injection values, and that in the p-region at $x = \infty$ $n_p = n_{po}$ and that in the n-region at $x = -\infty$ $p_n = p_{no}$, then the injected current density components can be obtained as

$$J_n = \frac{qD_n n_{po}}{L_n}\left[\exp\left(\frac{qV}{kT}\right) - 1\right] \tag{3.27}$$

$$J_p = \frac{qD_p p_{no}}{L_p}\left[\exp\left(\frac{qV}{kT}\right) - 1\right] \tag{3.28}$$

where L_n and L_p are the Debye lengths for electrons and holes respectively. The total carrier current density is therefore given by

$$J = J_n + J_p = \left(\frac{qD_n n_{po}}{L_n} + \frac{qD_p p_{no}}{L_p}\right)\left[\exp\left(\frac{qV}{kT}\right) - 1\right] = J_s\left[\exp\left(\frac{qV}{kT}\right) - 1\right] \tag{3.29}$$

A graph of current I against applied voltage V yields the familiar $I-V$ characteristic of the junction diode, Figure 3.3.

The closed-form analysis can be extended to derive incremental resistance, depletion and junction capacitance values. The incremental resistance, derived from the static charactersitics is,

$$r = \frac{dV}{dI} = \frac{kT}{qAJ_s[\exp(qV/kT)]} \tag{3.30}$$

where A is the cross-sectional area of the diode. At low forward bias voltages and in reverse bias, generation-recombination current dominates over the diffusion current and the incremental resistance is modified according to the relationship,

$$r = \frac{mkT}{qAJ} \tag{3.31}$$

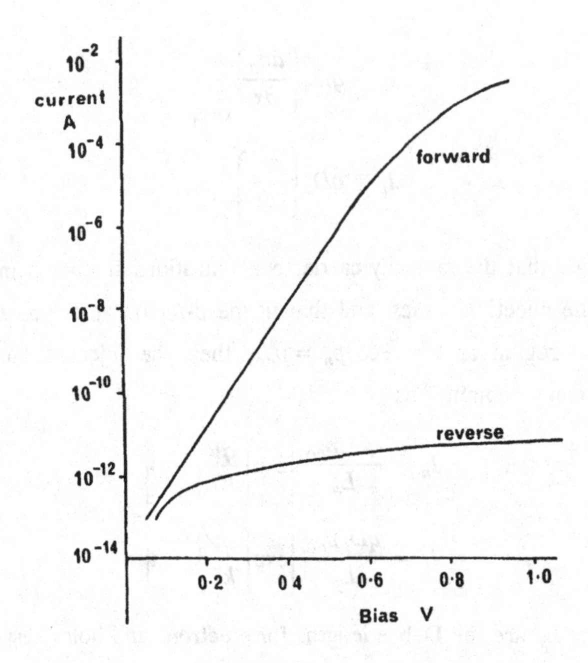

Figure 3.3 Current-voltage characteristics for pn junction

which may be derived from the Sah-Noyce-Shockley generation-recombination model [1]. The factor m allows for the bias dependence of the diode characteristic and for the differences in electrical behaviour between different types of diode. It is frequently taken as $m = 2$.

The presence of a charge dipole in the depletion region of the diode leads to a junction capacitance. The charge in the dipole is a function of the applied voltage. The depletion capacitance C_j may be obtained from,

$$C_j = \frac{dq}{dV} \tag{3.32}$$

The charge associated with each half of the depletion region is given by,

$$Q = q \int_{0}^{x_n} N_D(x)\, dx = q \int_{-x_p}^{0} N_A(x) dx \quad \textit{per unit area} \tag{3.33}$$

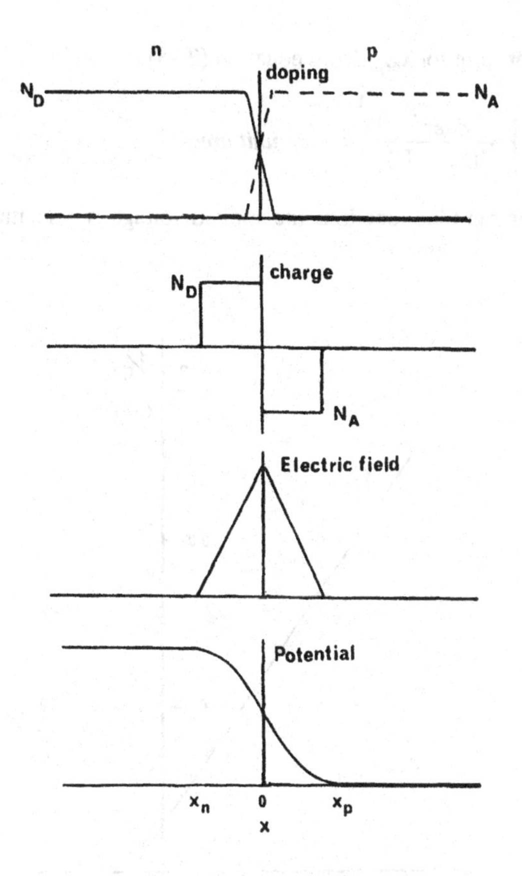

Figure 3.4 Abrupt pn junction - charge, electric field and potential profiles.

For an abrupt junction,

$$Q = qN_Dx_n = qN_Ax_p \quad \textit{per unit area} \tag{3.34}$$

which can be readily deduced by consideration of Figure 3.4. Hence substituting for x_n and x_p,

$$C_j = \left[\frac{q\varepsilon N_A N_D}{2(N_A + N_D)}\right]^{1/2} (V_j - V)^{-1/2} \quad \textit{per unit area} \tag{3.35}$$

The depletion capacitance for the abrupt junction follows the well known $1/C_d^2$ versus voltage relationship, Figure 3.5. In the case of a linearly graded junction it is necessary to substitute for x_{depl} from equation (3.18) to obtain,

$$C_j = \left(\frac{qa\varepsilon^2}{12(V_j - V)} \right)^{\frac{1}{3}} \quad \textit{per unit area} \tag{3.36}$$

Hence for linearly graded junctions the applied voltage is associtated with a $1/C_j^3$ dependence.

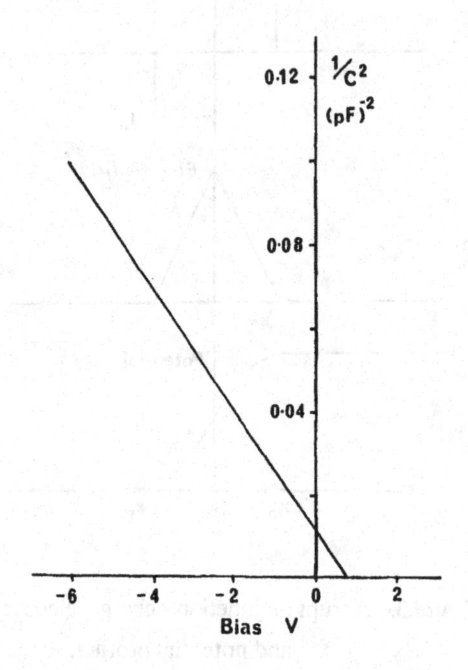

Figure 3.5 Capacitance-voltage characteristic for an abrupt pn junction

In forward bias minority carriers are injected into the neutral regions either side of the depletion region. This leads to excess carrier densities and associated additional stored charge Q_m. Since Q_m is a function of the applied voltage, a diffusion capacitance C_d can be defined,

$$C_d = \frac{dQ_m}{dV} \tag{3.37}$$

where

$$Q_m = q\left[\int_{-x_p}^{\infty} \Delta n_p\, dx + \int_{x_n}^{\infty} \Delta p_n\, dx\right] \quad per\ unit\ area \tag{3.38}$$

where Δn_p and Δp_n are the excess carrier densities. Using the same boundary conditions applied during the derivation of equations (3.27) and (3.28) and solving the continuity equations (3.23) and (3.24), the charge Q_m is extracted as,

$$Q_m = q(n_{po}L_n + p_{no}L_p)\left[\exp\left(\frac{qV}{kT}\right) - 1\right] \tag{3.39}$$

The diffusion capacitance is then obtained as,

$$C_d = \frac{q^2}{kT}(n_{po}L_n + p_{no}L_p)\exp\left(\frac{qV}{kT}\right) \quad per\ unit\ area \tag{3.40}$$

The total junction capacitance is the sum of the depletion and diffusion capacitances. At low forward bias voltages and in reverse bias, the depletion capacitance is the dominant component. In regions of moderate forward bias the diffusion component is the more significant capacitance, for voltages below V_j. The depletion and diffusion capacitances derived above are based on the static voltage-current characteristics and are usually termed 'quasi-static'. These expressions only strictly apply at low frequencies and do not reflect the frequency dependence of these parameters, which is significant at higher frequencies.

Junction breakdown in pn junctions occurs as a result of the very high electric field present across the depletion region under conditions of high reverse bias. The two mechanisms involved in the breakdown process are zener breakdown and avalanche breakdown. Zener breakdown is due to quantum mechanical tunnelling, where carriers with sufficient energy tunnel through the barrier. The current density due to quantum mechanical tunneling may be expressed as [2],

$$J_z = \frac{q^3(2m^*)^{1/2}}{4\pi^2 h^2 \xi_g^{1/2}} \exp\left[-\frac{4\xi_g^{\frac{3}{2}}(2m^*)^{1/2}}{3q\hbar E}\right] EV \tag{3.41}$$

where ξ_g is the energy band gap and m^* is the effective mass. The electric field must approach 100 MVm^{-1} before the tunnelling current becomes significant. Avalanche breakdown is due to avalanche multiplication of carriers injected into the depletion region, in the presence of very high electric fields. The avalanche breakdown voltage V_B can be obtained for known depletion boundaries and an abrupt junction as,

$$V_B = \frac{E_M x_{depl}}{2} = \frac{\varepsilon E_M^2 (N_A + N_D)}{2qN_A N_D} \tag{3.42}$$

For the case where one side of the junction is doped more lightly than the other side,

$$V_B = \frac{\varepsilon E_M^2}{2qN_D} \quad \text{for } N_A \gg N_D \tag{3.43}$$

$$V_B = \frac{\varepsilon E_M^2}{2qN_A} \quad \text{for } N_D \gg N_A \tag{3.44}$$

For linearly graded junctions,

$$V_B = \frac{2E_M x_{depl}}{3} = \frac{4E_M^{\frac{3}{2}}}{3}\left(\frac{2\varepsilon}{qa}\right)^{1/2} \tag{3.45}$$

The closed-form experessions derived above for the electrical characteristics of pn junctions may be applied to simplified analyses of other bipolar devices in addition to the simple pn junction diode. In particular they provide a useful starting point for analysing the operation of bipolar junction transistors.

3.2.2. Analysis of Field Effect Transistor Operation

A starting point for many closed-form field effect transistor analyses is the classical 'Gradual Channel' method described by Shockley in 1952 [3]. The gradual channel approximation assumes that the device is symmetrical and that the conducting channel is restricted to the active channel region. An abrupt interface is assumed between the active channel and the depletion region and diffusion is neglected. This implies that the carrier concentration n is zero in the depletion region and equal to the doping density ($n = N_D$) in the active channel. The gradual channel model assumes that changes along the channel are slow compared with changes across the channel ($d^2V/dx^2 = 0$). This latter assumption is hard to justify for short gate length devices with small aspect ratios (gate length to active channel thickness ratio). The analysis assumes that for values of drain voltage beyond pinch-off the drain current remains constatnt at a saturated value. The analysis was originally applied to junction field effect transistors (JFET's) and has subsequently been applied to Schottky barrier gate field effect transistors (MESFET's). An n-type channel is considered here using a majority carrier analysis (the FET is a unipolar device).

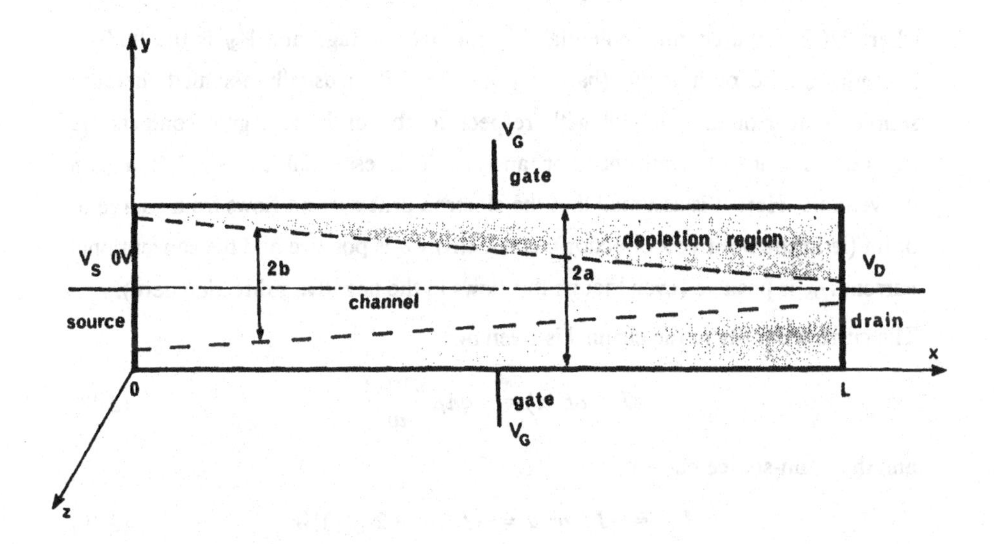

Figure 3.6 Field effect transistor model (symmetrical case)

The analysis concentrates on the active channel region below the gate, although the model can be easily extended to include the neutral source and drain regions which contribute to the overall channel resistance (parasitic resistances R_S and R_D respectively). The FET model is shown in Figure 3.6, where the device is assumed symmetrical and the channel length is taken as the direction of current flow. It is important to distinguish between the gate length L_G (the distance between the source edge and the drain edge of the gate) and the gate width W (the dimension parallel to the drain and source contacts) - usually the gate width is much larger than the gate length. The active channel thickness before the depletion layer thickness is taken into account is defined as $2A$ (symmetrical about the mid point of the channel).

The undepleted channel height is given by

$$h = 2A - 2y_d(x) \tag{3.46}$$

where $y_d(x)$ is the thickness of the depletion region given by

$$y_d(x) = \left[\frac{2\varepsilon_r\varepsilon_o(V(x)+V_{Bi}-V_G)}{qN_D}\right]^{1/2} \tag{3.47}$$

where V(x) is the channel potential, V_G the gate voltage and V_{Bi} is the built-in potential of the pn-junction (barrier potential). It is usually assumed that the source is at ground potential with respect to the drain and gate contacts (ie $V_S = 0$). As in all semiconductor analysis it is essential to establish a sign convention. Here it is assumed that the electron drift current flows from source to drain (ie in the positive x direction), conductivity σ is positive and the conventional current I_{DS} is positive (even though this flows in the negative x direction here). The current density in the channel is given by

$$J = \sigma E(x) = -qn\mu\frac{dV(x)}{dx} \tag{3.48}$$

and the drain-source current,

$$I_{DS} = -J \times area = -J(2A - 2y_d(x))W \tag{3.49}$$

The negative sign is inserted to preserve the sign convention and yield positive values of I_{DS}. Substituting for the current density J, depletion thickness y_d and assuming $n = N_D$,

$$I_{DS} = 2qN_D\mu W\left\{A - \left[\frac{2\varepsilon_r\varepsilon_o(V(x)+V_{Bi}-V_G)}{qN_D}\right]^{1/2}\right\}\frac{dV(x)}{dx} \tag{3.50}$$

Separating variables and integrating along the channel, yields

$$I_{DS} = G_0\left\{V_D - \frac{2}{3}\left(\frac{2\varepsilon_r\varepsilon_o}{qN_DA^2}\right)^{1/2}\left[(V_D + V_{Bi} - V_G)^{\frac{3}{2}} - (V_{Bi}-V_G)^{\frac{3}{2}}\right]\right\} \tag{3.51}$$

where the zero current ($V_D = 0$) channel conductance G_0 is defined as,

$$G_0 = \frac{2qN_D\mu WA}{L_G} \tag{3.52}$$

The channel becomes 'pinched-off' when the reverse-bias across the depletion layer ($V_G - V(x)$) reaches a value V_P, called the pinch-off voltage. Under these conditions the channel is fully depleted from the point x onwards (ie. $y_d = A$). The pinch-off voltage is given as,

$$V_P = \frac{qN_D A^2}{2\varepsilon_r \varepsilon_o} \tag{3.53}$$

This expression is frequently referred to when characterising FET's. hence the current may be defined as

$$I_{DS} = G_0 \left\{ V_D - \frac{2}{3} \left[\frac{(V_D + V_{Bi} - V_G)^{\frac{3}{2}} - (V_{Bi} - V_G)}{V_P^{1/2}} \right] \right\} \tag{3.54}$$

In the case of an N-channel FET, V_G and V_P are negative and V_D is positive. Equation (3.54) which is often referred to as the 'fundamental equation' of field effect transistors, describes the drain current and is only valid for the linear region below pinch-off (where $|V_G - V_D| < |V_P|$). This condition allows the characteristics to the left of the dotted line in Figure 3.7 to be obtained. Beyond pinch-off the drain current is assumed to remain constant at the value attained at the onset of 'saturation' (pinch-off). Th corresponding value of drain to source voltage is known as the saturation voltage $V_{DS(sat)}$.

An important parameter which can be derived from the current-voltage characteristics is the transconductance g_m, which is a measure of the gain of the transistor, and is defined as

$$g_m = \left. \frac{\partial I_{DS}}{\partial V_{GS}} \right|_{V_{DS} = constant} \tag{3.55}$$

hence

$$g_m = G_0 \frac{(V_D + V_{Bi} - V_G)^{1/2} - (V_{Bi} - V_G)^{1/2}}{V_P^{1/2}} \tag{3.56}$$

The drain (output) conductance g_d is similarly defined as

$$g_d = \left. \frac{\partial I_{DS}}{\partial V_D} \right|_{V_G = constant} \tag{3.57}$$

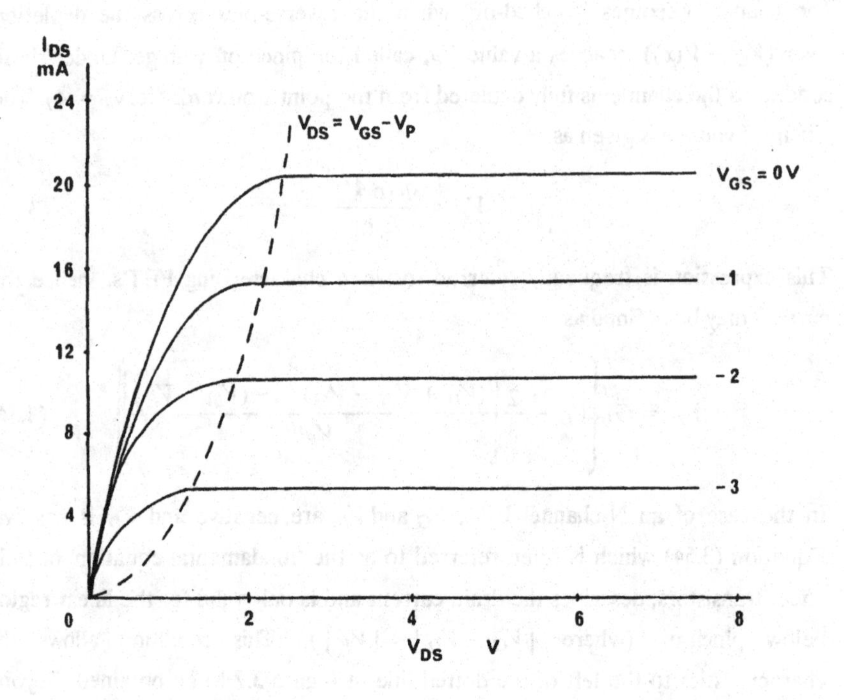

Figure 3.7 FET characteristics obtained from one-dimensional model

hence

$$g_d = G_0\left[1 - \left(\frac{V_D + V_{Bi} - V_G}{V_P}\right)^{1/2}\right] \tag{3.58}$$

The transconductance and output resistance are usually evaluated in the saturation regime, where the device is normally operated. Hence assuming a constant current equal to that at pinch-off, and substituting $V_{DS}(sat) = V_G - V_P$, the transconductance and drain conductance for the saturation region can be approximated to

$$g_m = G_0\left[1 - \left(\frac{V_G}{V_P}\right)^{1/2}\right] \tag{3.59}$$

$$g_d \simeq 0 \tag{3.60}$$

where the saturated drain to source current I_{DS} is given by,

$$I_{DS(sat)} = G_0 \left\{ \frac{2}{3} \left[\left(\frac{V_{Bi} - V_G}{V_P} \right)^{1/2} - 1 \right] (V_{Bi} - V_G) + \frac{1}{3}(V_G - V_{Bi} - V_P) \right\} \tag{3.61}$$

The maximum current I_{DSS} at $V_G = 0$ is given by,

$$I_{DSS} = \frac{WA^3 \mu q^2 N_D^2}{6\epsilon L_G} \tag{3.62}$$

The approximation for the saturation region leads to zero output conductance and constant drain current for a given value of gate voltage V_G. This can be compared with the finite drain conductance (typically $g_d = 25\mu S$ for a silicon N-channel JFET) and empirical square law saturation current characteristic found in practical devices where,

$$I_{DS(sat)} = I_{DSS} \left(1 - \frac{V_G}{V_P} \right)^2 \tag{3.63}$$

The basic analysis presented above has been developed by several authors to account for drift velocity saturation effects in both silicon and gallium arsenide FET's, and the presence of dipole regions in GaAs devices [4,5,6] presents an extension of Shockley's basic theory, applied to GaAs MESFET's, which attempts to account for drift velocity saturation effects. This approach again assumes a constant saturation current and includes a treatment of domain formation at saturation. Shur assumes that for equation (38) to remain valid,

$$V_i \leq V_{sat} \quad where \quad V_{sat} = E_s L_G \tag{3.64}$$

V_i is the channel voltage, V_{sat} is the channel voltage corresponding to the saturation electron drift velocity and E_s is the domain sustaining field [6].

More sophisticated models based on two-dimensional field approximations have been developed [7,8,9]. Many of these two-dimensional models sub-divide the active channel into separate regions with different approximations applied to each region. For example, Lehovec and Zuleeg [10], included a two-dimensional field approximation between the gate and drain. Their technique was based upon a solution obtained by superimposing the solution for Laplace's equation on the field solutions obtained for Poisson's equation. The channel region in which the carrier concentration remains approximately constant, where the carrier velocity is below saturation, is modelled using the gradual channel relationships described above.

Techniques which combine the advantages of rapidly evaluated analytical approximations with the accuracy and flexibility of numerical schemes have been used to model MESFET's [8,9]. These models include more detailed drift-velocity characteristics, and account for a gradual depletion boundary (in contrast to the abrupt transition assumed in the analysis above). The channel below the gate is sub-divided into three regions, comprising the neutral channel, a transition region and a fully depleted region. The Poisson and Laplace equations are solved over these regions so that the general potential solution ψ is given by

$$\psi = \psi_o + \psi_1 \tag{3.65}$$

where

$$\nabla^2\psi_0 = 0 \quad \text{and} \quad \nabla^2\psi_1 = -\frac{q}{\varepsilon_r\varepsilon_o}[N_D - n(x,y)] \tag{3.66}$$

The solution for ψ_1 extends the depletion region into the channel. The full analysis is rather lengthy and is dealt with in [8] and [9]. The process of determining the correct potential solution requires an iterative technique to extract ψ_0 at the limits of the gate $x = 0$ and $x = L_G$.

3.2.3. Analysis of MOSFET Operation

Another type of transistor which can be analysed to yield simplified closed-form expressions is the metal-oxide semiconductor field effect transistor (MOSFET). A simplified N-channel MOSFET geometry often used in association with closed-form analyses is shown in Figure 3.8. The following assumptions are usually adopted:

(i) The potential gradient in the channel is much smaller than that due to the gate in the perpendicular direction.

(ii) A constant average electron mobility μ is assumed in the inverted channel region. The value is usually taken as half the bulk mobility value, because of increased carrier scattering at the $Si - SiO_2$ inteface.

(iii) The source is assumed to be externally connected to the substrate.

(iv) It is assumed that the threshold voltage V_T is constant with distance x along the inverted channel. This approximation becomes increasingly inaccurate with increasing substrate doping level.

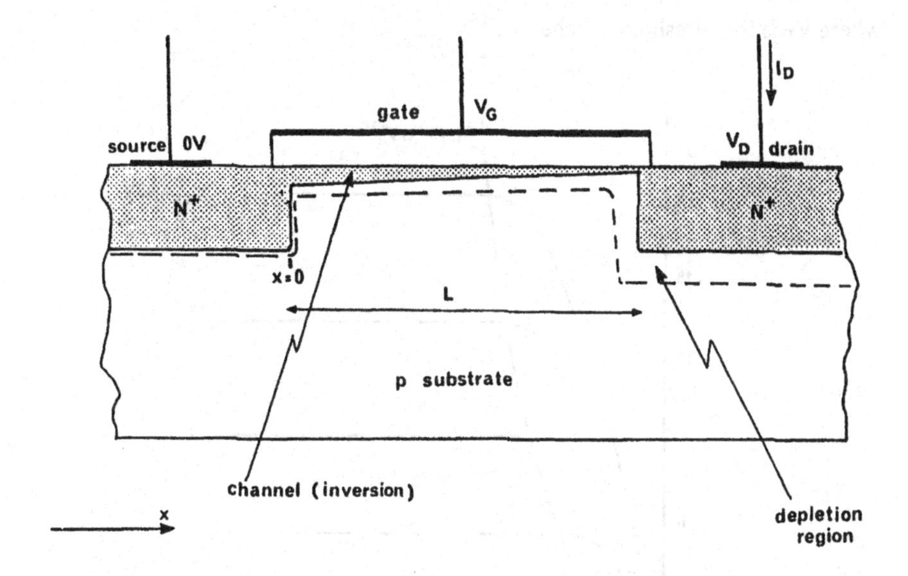

Figure 3.8 Cross-section of a n-channel MOSFET

The same sign convention as adopted previously will be applied here. The drain-source current (conventional current) is given by

$$I_{DS} = q\mu WQ(x)\frac{dV(x)}{dx} \tag{3.67}$$

where the electronic charge per unit area in the channel (minority carriers in the substrate) as a function of distance x is given by,

$$Q(x) = C_{ox}[(V_{GS} - V_T) - V(x)] \tag{3.68}$$

where C_{ox} is the oxide capacitance and is given by

$$C_{ox} = -\frac{dQ_s}{dVox} = \frac{\varepsilon_{ox}\varepsilon_o}{t_{ox}} \tag{3.69}$$

where Q_s is the charge in the semiconductor, V_{ox} is the voltage dropped across the oxide layer and t_{ox} is the thickness of the oxide. Substituting for Q(x), separating the variables and integrating equation (49) yields,

$$I_{DS} = \frac{\mu C_{ox} W}{L_G}\left[(V_{GS} - V_T)V_{DS} - \frac{V_{DS}^2}{2}\right] \quad \text{for} \quad 0 \leq V_{DS} \leq (V_{GS} - V_T) \qquad (3.70)$$

where V_T is the threshold voltage.

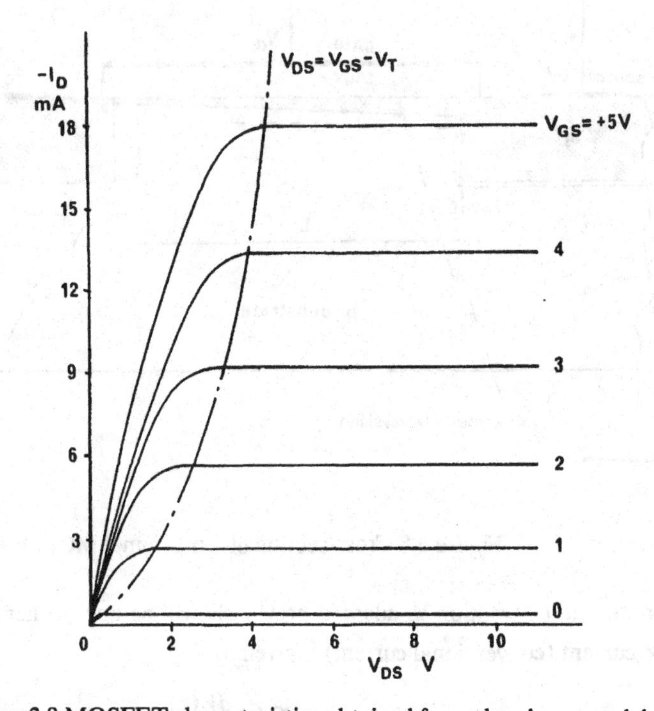

Figure 3.9 MOSFET characteristics obtained from the above model

For the condition $V_{DS} = V_{GS} - V_G$, the gate-to-channel voltage at the drain end of the channel is just sufficient to maintain inversion and Q(x) vanishes. It would appear that under these conditions the field would have to become infinite according to equation (49), which is clearly impossible. In practice the first of the specified assumptions that dV/dx is negligible is no longer valid. Furthermore, the carrier drift velocity reaches a saturated value. Hence, channel pinch-off and current saturation occur at $V_{DS} = V_{GS} - V_T$, and for applied voltages greater than this, the field an the pinched-off region increases without a significant further increase in current. This is because the carriers constituting the mobile charge Q(x) are already moving at the saturated velocity. Hence in saturation,

$$I_{DS} = \frac{\mu C_{ox} W}{2L}(V_{GS} - V_T)^2 \quad \text{for } V_{DS} \geq V_{GS} - V_T \tag{3.71}$$

An example of the type of characteristics predicted by this analysis is shown in Figure 3.9. The transconductance can be shown to be

$$g_m = \frac{\mu C_{ox} W}{L}(V_{GS} - V_T) \tag{3.72}$$

A more detailed discussion of the operation of JFET's, MOSFET's and MESFET's can be found in general semiconductor texts such as [10,11,12,13].

3.3. Limitations of Closed-Form Analyses

In practice, in all but the very simplest of devices, analysis techniques which yield closed-form expressions for the terminal charactersitics, internal field and carrier distributions all require approximations to allow the carrier transport equations to be solved. One-dimensional device structures, such as vertical diodes and long gate length FET's, may be analysed to yield solutions which provide useful approximations to device charactersitics. However, this type of analytical method is severly limited in its accuracy for describing two-dimensional structures, such as planar transistors, or small-scale devices where fringing fields and other multi-dimensional effects predominate.

An important feature of closed-form analytical techniques is that they provide a rapid means of estimating the charactersitics of many devices, even if the true charactersitics depart from those predicted by the simplified models. In circumstances where greater accuracy and flexibility are required it is necessary to resort to full numerical solutions of the complete carrier transport equations.

References

[1] Sah, C.T., Noyce, R.N. and Shockley, W., "Carrier generation and recombination in p-n junctions and p-n junction characteristics", Proc. IRE, 45, pp1228- , 1957

[2] Moll, J.L., "Physics of Semiconductors", McGraw-Hill, New York, 1964

[3] Shockley, W., "A unipolar 'field-effect' transistor", Proc. IRE, pp.1365-1377, 1952

[4] Dacey, G.C. and Ross, I.M., "The field effect transistor", Bell Syst. Tech. J., Vol.34, p1149, 1955

[5] Hauser, J.R., "Characteristics of junction field effect devices with small channel length-to-width ratios", Solid State Electron., Vol.10, pp.577-587, 1967

[6] Shur, M., "Analytical model of GaAs MESFET's", IEEE Trans. Electron Devices, Vol. ED-25, No.6, pp.612-618, 1978

[7] Turner, J.A and Wilson, B.L.H., "Implications of carrier velocity saturation in a gallium arsenide field effect transistor", Proc. Gallium Arsenide Inst. Phys. Conf. Ser. No.7, ed H Strack (Bristol, Institute of Physics), p195, 1968

[8] Shur, M.S. and Eastman, L.F., "Current-voltage characteristics, small signal parameters, and switching times of GaAs FETs", IEEE Trans. Electron Devices, Vol.ED-25, No.6, pp.601-611, 1978

[9] Yamaguchi, K. and Kodera, H., "Drain conductance of junction gate FETs in the hot electron range", IEEE Trans. Electron Devices, Vol.ED-23, No.6, pp.545-553, 1976

[10] Lehovec, K. and Zuleeg, R., "Voltage-current characteristics of GaAs J-FETs in the hot electron range", Solid State Electron., Vol.13, pp.1415-1426, 1970

[11] Bar-Lev, A., "Semiconductors and Electronic Devices", London, Prentice-Hall International, 1979

[12] Sze, S.M., "Physics of Semiconductor Devices", Wiley International, 1969

[13] Dascalu, D., "Electronic processes in unipolar solid-state devices", Abacus Press (UK), 1977

CHAPTER 4

NUMERICAL SOLUTION OF THE SEMICONDUCTOR EQUATIONS THE FINITE-DIFFERENCE METHOD

In general it is not possible to obtain closed-form expressions which describe satisfactorily the operation of modern semiconductor devices. The degree of precision and operational parameters associated with analytic models are usually limited by the approximations necessary to obtain the closed-form description. In contrast, numerical techniques can be used to solve the full set of semiconductor equations over a specified domain. However, the numerical algorithms and computer resources required to solve the partial differential equations, which describe the carrier transport process, can in some circumstances be quite formidable. Despite this, simulation packages are becoming available which can even be run satifactorily on the more powerful desk-top computers.

The most common numerical techniques used to solve the set of partial differential equations which constitute the semiconductor equations, are the finite-difference and finite-element methods. Both methods rely on the discretization of the equations across the specified geometrical domain of the device model. Other techniques which have been successfully applied to the problem include the boundary integral method [1,2].

4.1. Finite-Difference Schemes

The finite-difference method is particularly well suited to simple device geometries and has been used extensively to model one-dimensional and two-dimensional rectangular geometry (planar) devices. More recently three-dimensional models have been developed using the relatively easily implemented algorithms. Finite-difference techniques applied to semiconductor modelling are well established and there is considerable information available on the stability and convergence properties of these schemes (see for example [3,4,5]). The majority of earlier device simulations such as those of Gummel [6] Slotboom [7] and Scharfetter and Gummel [8] used one-dimensional finite-difference schemes. Subsequently, even the more sophisticated simulations of the early seventies [3,4] relied almost entirely on this method, even though other numerical solution techniques had become established. This was largely due to the additional complexity introduced into the simulation algorithms by alternative methods. Since that time the powerful finite-element method has become more popular for semiconductor simulations and is becoming as common as the finite-difference method. The recent introduction of new techniques, such as finite boxes, allows finite difference methods to be utilised in applications which would previously have been better suited to the finite-element method.

The finite-difference methods produce solutions for the physical variables ψ, n, p as discretized values at specific nodes contained within a mesh superimposed on the solution domain. The continous derivatives of the semiconductor equations are replaced by discretized finite-difference approximations. The discretized physical variables are represented by values obtained from the solution of the discretized equations at each mesh point except where boundary conditions determine the values of the variables.

The finite-difference equations are derived from truncated Taylor series for $f(x+\Delta x)$ and $f(x-\Delta x)$, and assume that the function $f(x)$ is a continuous, single valued function of x with continuous derivatives. First-order approximations are obtained from these Taylor expansions as forward or backward difference formulae respectively, by truncating the series after the second term. Second-order approximations may also be obtained by subtracting or adding the truncated Taylor series to yield first and second differentials respectively, which are also known as central-difference approximations and are on order of Δx more accurate than the first order expressions. The small incremental change Δx between mesh points, is constant in a uniform mesh and varies in a non-uniform mesh. Hence to maintain generality it is convenient to define the distance between adjacent mesh points using the notation,

$$a_i = \Delta x = x_{i+1} - x_i \qquad i = 0,1,2,3...n \tag{4.1}$$

$$b_j = \Delta y = y_{j+1} - y_j \qquad j = 0,1,2,3...m \tag{4.2}$$

In the following discussion of finite-difference techniques it will be assumed that a two-dimensional model is being considered. In describing two-dimensional finite-difference schemes it is convenient to use the notation

$$f(x,y) = f_{i,j} \qquad i = 0,1,2...n \quad j = 0,1,2...m \tag{4.3}$$

$$f(x + \Delta x, y) = f_{i+1,j} \tag{4.4}$$

$$f\left(\frac{x+\Delta x}{2}, y\right) = f_{i+1/2,j} \quad etc \tag{4.5}$$

where the relative positions of the points i,j etc are in the Cartesian plane. Note that it is assumed that Δx and Δy are variable in the above discussion and are defined by equations (4.1) and (4.2). This notation is used to describe a mesh superimposed on the region requiring analysis. The notation is illustrated in Figure 4.1, where the mesh points and 'half-points' are clearly marked. These mesh points constitute the well known 'five-point molecule', which consists of the central node and the surrounding four points.

Uniform meshes are frequently used because of the ease of calculation, Figure 4.2. Non-uniform meshes can be advantageous in regions of high and low valued derivatives, with fine and course grid spacing respectively, Figure 4.3. This allows greater accuracy for the higher derivatives and a saving in time and memory in regions of low valued derivatives [4]. A major disadvantage of the finite-difference technique when compared with the finite-element method, is that the requirement for a finer mesh in one specific area of the device necessitates the presence of a finer mesh in other parts of the device (see Figure 4.3). This leads to a surplus of nodes in regions away from the region requiring refinement which would otherwise have a relatively course mesh. The recently reported finite box method of Franz et al [9] avoids this situation and allows finite-difference schemes to be applied to more complex non-planar geometries. The finite box method allows mesh lines to terminate within the mesh, on mesh lines normal to the direction of the incident (terminating) line, Figure 4.4. This produces a considerable saving in array size and computation time. The finite boxes technique generates and adapts

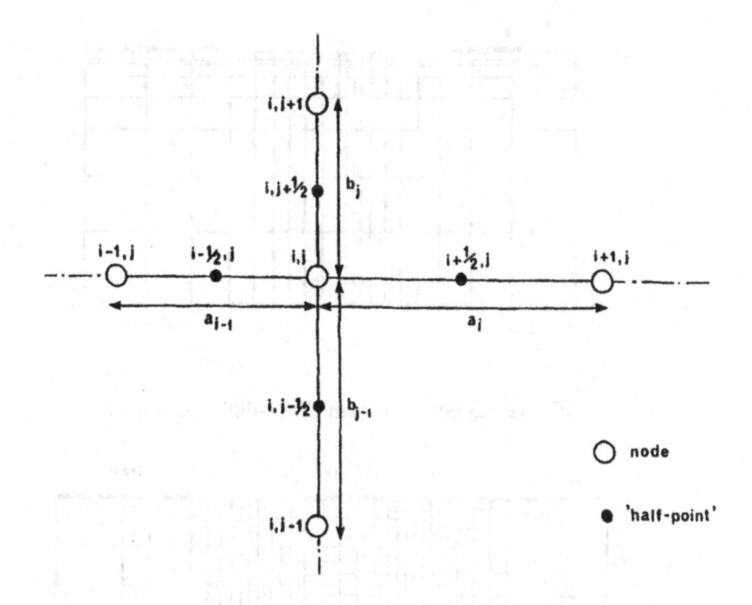

Figure 4.1 Finite-difference mesh point notation

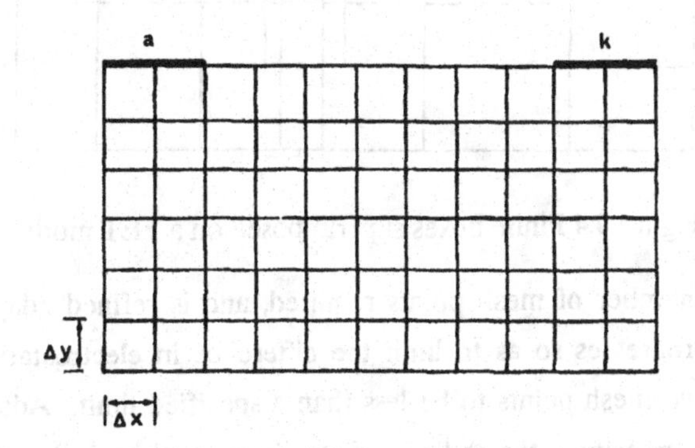

Figure 4.2 Uniform finite-difference mesh

the grid automatically.

The accuracy and efficiency of finite-difference schemes can be substancially improved by using adaptive meshes. This type of mesh is 'self generating' and produces localised mesh refinement to take into account changes in the potential and carrier distributions. The mesh is chosen so as to minimise the discretization

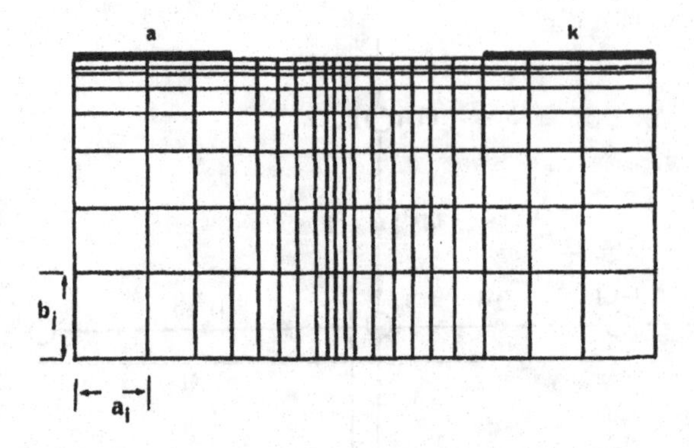

Figure 4.3 Non-uniform finite-difference mesh

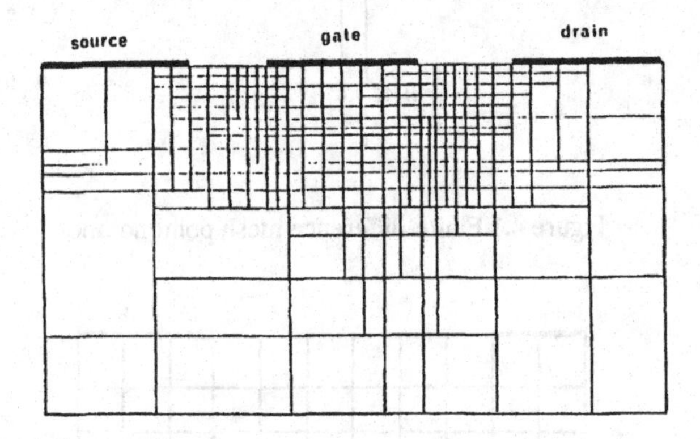

Figure 4.4 Finite boxes superimposed on a FET model

error and the number of mesh points required, and is refined adaptively as the computation progresses so as to limit the difference in electrostatic potential ψ between any two mesh points to be less than a specified limit. Adaptive meshes increase the complexity of the coding and may increase the solution time required, but provide a considerable improvement in accuracy and stability.

4.2. Discretization of the Semiconductor Equations

The Poisson equation is usually discretized using a 'five-point' difference approximation, where the potential at a node depends on the potential and charge at the node and on the potential at the four neighbouring nodes. The finite-difference equation for the Poisson equation over a uniform mesh is thus,

$$\frac{\psi_{i-1,j} - 2\psi_{i,j} + \psi_{i+1,j}}{\Delta x^2} + \frac{\psi_{i,j-1} - 2\psi_{i,j} + \psi_{i,j+1}}{\Delta y^2}$$

$$= -\frac{q}{\varepsilon_o \varepsilon_r}\left(N_{D_{i,j}} - n_{i,j} + p_{i,j} - N_{A_{i,j}}\right) \tag{4.6}$$

The error in this discretised form of the Poisson equation is of the order of $(\Delta x^2, \Delta y^2)$. In the case of a non-uniform mesh the finite-difference equation becomes (using the notation of equations (4.1) and (4.2)),

$$\frac{\dfrac{\psi_{i+1,j} - \psi_{i,j}}{a_i} - \dfrac{\psi_{i,j} - \psi_{i-1,j}}{a_{i-1}}}{\dfrac{a_i + a_{i-1}}{2}} + \frac{\dfrac{\psi_{i,j+1} - \psi_{i,j}}{b_j} - \dfrac{\psi_{i,j} - \psi_{i,j-1}}{b_{j-1}}}{\dfrac{b_j + b_{j-1}}{2}}$$

$$= -\frac{q}{\varepsilon_o \varepsilon_r}\left(N_{D_{i,j}} - n_{i,j} + p_{i,j} - N_{A_{i,j}}\right) \tag{4.7}$$

The discretised Poisson may be expressed in matrix form as

$$[A][\psi] = [B] \tag{4.8}$$

where matrix $[A]$ is the coefficient matrix, normally of tridiagonal form and matrix $[B]$ contains the terms on the right hand side of equation (4.7). The coefficient matrix normally has a maximum of five non-zero elements on any row of the matrix and is therefore termed 'sparse'.

The electric field $\mathbf{E}$ is conveniently resolved into two components E_x and E_y in the x and y directions respectively, so that

$$E = \sqrt{(E_x^2 + E_y^2)} \tag{4.9}$$

The components E_x and E_y can be described in finite-difference form using forward (or backward) difference or central difference notation. For the case of forward difference notation

$$E_{x_{i,j}} = -\frac{\partial \psi}{\partial x} = \frac{\psi_{i,j} - \psi_{i+1,j}}{a_i} + O(a_i) \tag{4.10}$$

$$E_{y_{i,j}} = -\frac{\partial \psi}{\partial y} = \frac{\psi_{i,j} - \psi_{i,j+1}}{b_j} + O(b_j) \tag{4.11}$$

Greater accuracy is achieved using central difference schemes, where in the case of a uniform mesh,

$$E_{x_{i,j}} = -\frac{\partial\psi}{\partial x} = \frac{\psi_{i-1,j} - \psi_{i+1,j}}{2\Delta x} + O(\Delta x^2) \tag{4.12}$$

$$E_{y_{i,j}} = -\frac{\partial\psi}{\partial y} = \frac{\psi_{i,j-1} - \psi_{i,j+1}}{2\Delta x} + O(\Delta x^2) \tag{4.13}$$

It should be noted that if a 'half-point' notation is chosen, then equations (4.10) and (4.11) correspond to central difference schemes for expressing $\partial\psi/\partial x$ at $(i+1/2,j)$ and $\partial\psi/\partial y$ at $(i,j+1/2)$, and have an improved accuracy of order $O(a^2)$ and $O(b^2)$ respectively. The electric field E is used to determine the mobility μ and diffusion D in classical semiconductor simulations. The mobility and diffusion coefficients are usually treated as scalar functions of the magnitude of the local electric field E.

The finite-difference discretization of the current continuity equation may be carried out using a number of different forms [4,6,8,10]. In contrast to the elliptic Poisson equation, the solution of the parabolic continuity equations can present some difficulties due to the time-dependence and non-linear velocity-field relationship. There are two principal approaches to discretizing the continuity equation - schemes utilising the explicit current densities at the nodes and schemes based on eliminating the current densities from the continuity equations by substituting for J_x and J_y and subsequently discretising the resulting equations. This latter approach works well in devices where localised rapid changes in carrier concentration have little effect on the bulk properties of the device, such as in transferred electron devices, or for materials with small diffusion coefficients like silicon [11,12]. A typical expansion of the continuity equation is

$$\frac{\partial n}{\partial t} = n\nabla(\mu\mathbf{E}) + \mu\mathbf{E}.\nabla n + D\nabla^2 n + \nabla n.\nabla D \tag{4.14}$$

The finite-difference equation is then obtained by substituting standard finite-difference expansions for the derivatives. Such schemes usually require less computation than the former method. However, for devices with high carrier concentrations ($>10^{23}m^{-3}$), or for materials with large diffusion coefficients (gallium arsenide, indium phosphide), the truncation error in higher order derivative terms can become significant, particularly in the vicinity of depletion regions. The most useful approach is to develop a finite-difference scheme based on the expansions utilizing current densities at each node [3,4,6].

The finite-difference formulation of the continuity scheme usually follows a 'half-point' difference expansion based on central difference approximations. For example the current continuity equation for electrons in a majority carrier device could be written,

$$\frac{\partial n}{\partial t} = \frac{1}{q}\nabla.\mathbf{J} = \frac{1}{q}\left(\frac{J_{x_{i+1/2},j} - J_{x_{i-1/2},j}}{\Delta x} + \frac{J_{y_{i,j+1/2}} - J_{y_{i,j-1/2}}}{\Delta y}\right)$$

$$+ O(\Delta x^2, \Delta y^2, \Delta t) \tag{4.15}$$

or for the case of a non-uniform mesh,

$$\frac{\partial n}{\partial t} = \frac{1}{q}\nabla.\mathbf{J} = \frac{1}{q}\left(\frac{J_{x_{i+1/2},j} - J_{x_{i-1/2},j}}{\frac{a_i + a_{i-1}}{2}} + \frac{J_{y_{i,j+1/2}} - J_{y_{i,j-1/2}}}{\frac{b_j + b_{j-1}}{2}}\right)$$

$$+ O(a,b,\Delta t) \tag{4.16}$$

The corresponding electron current densities in this expression may be obtained by substitution of discretized values and difference expansions into the conventional current density equations,

$$J_{x_{i+1/2},j} = qn_{i+1/2,j}\mu_{i+1/2,j}E_{x_{i+1/2},j} + qD_{i+1/2,j}\frac{n_{i+1,j} - n_{i,j}}{a_i} \tag{4.17}$$

$$J_{x_{i-1/2},j} = qn_{i-1/2,j}\mu_{i-1/2,j}E_{x_{i-1/2},j} + qD_{i-1/2,j}\frac{n_{i,j} - n_{i-1,j}}{a_{i-1}} \tag{4.18}$$

$$J_{y_{i,j+1/2}} = qn_{i,j+1/2}\mu_{i,j+1/2}E_{y_{i,j+1/2}} + qD_{i,j+1/2}\frac{n_{i,j+1} - n_{i,j}}{b_j} \tag{4.19}$$

$$J_{y_{i,j-1/2}} = qn_{i,j-1/2}\mu_{i,j-1/2}E_{y_{i,j-1/2}} + qD_{i,j-1/2}\frac{n_{i,j} - n_{i,j-1}}{b_{j-1}} \tag{4.20}$$

The value of any function f (n, μ, E and D above) at the half-point is given by the linear average of the function at mesh points either side of the half-point,

$$f_{i+1/2,j} = \frac{f_{i+1,j} + f_{i,j}}{2} \tag{4.21}$$

$$f_{i,j+1/2} = \frac{f_{i,j+1} + f_{i,j}}{2} \qquad (4.22)$$

The electric fields $E_{x_{i+1/2,j}}$ etc are given by the expression,

$$E_{x_{i+1/2,j}} = -\left.\frac{\partial\psi}{\partial x}\right|_{i+1/2,j} = -\frac{\psi_{i+1,j} - \psi_{i,j}}{a_i} + O(a_i^2) \;\; etc \qquad (4.23)$$

Explicit schemes usually require excessively small time steps to guarantee stability and obtain accurate solutions. A Crank-Nicolson implicit scheme is usually adopted, which provides greater accuracy and better convergence properties than is possible with explicit solution methods. The general implicit form of the continuity equation (for electrons) based on current densities is

$$\frac{\partial n}{\partial t} = \frac{n_{i,j}^{k+1} - n_{i,j}}{\Delta t} = \frac{1}{2q}\left[\nabla.\mathbf{J}_{i,j}^{k} + \nabla.\mathbf{J}_{i,j}^{k+1}\right] \qquad (4.24)$$

where the superscript k refers to the time step. If this equation is treated as being fully implicit in both velocity (μE) and carrier concentration, it leads to a complicated system of coupled, non-linear equations, which is difficult to solve. An advantage of this approach is that it is stable for all values of time and space step which contrasts with explicit schemes which have strong tendencies to instability. However, smaller values of time step Δt, Δx and Δy yield more accurate results. The continuity scheme is usually linearised to allow conventional methods of solving linear systems of equations to be used [13]. The linearisation usually implemented requires that,

$$\mathbf{v}_{i,j}^{k+1} = \mathbf{v}_{i,j}^{k} \qquad (4.25)$$

which makes the equation implicit in carrier concentration and explicit in velocity. This leads to an asymmetric coefficient matrix which yields complex eigenvalues in the Jacobi iteration matrix. As a result of this successive under relaxation (SUR) has to be used to solve this set of equations rather than the more familiar successive over relaxation (SOR) technique.

Stability analysis for non-linear systems is a highly complicated process and no generalised analysis is available. Reiser has shown that for linear explicit solutions of the continuity equation [14] the time step is restricted for stability reasons by conditions given by,

$$\Delta t < \min\left[\frac{\Delta x^2 \Delta y^2}{2D(\Delta x^2 + \Delta y^2)}, \frac{2D}{v_\infty^2}\right] \tag{4.26}$$

where v_∞ is the high field saturated constant velocity. Richtmeyer and Marton indicate that the approximations obtained using linear stability analysis theory also provide useful guide lines for non-linear problems [5]. However, the explicit treatment of drift velocity in the linearised continuity equation gives rise to non-linear instabilities which manifest themselves as standing waves superimposed on the required solution when large time steps are used [14]. Furthermore, these erroneous perturbations are amplified rather than damped if the step width is excessive and it is necessary to restrict the time steps to less than 1 *ps* to minimise this effect. In order to maintain a physically meaningful solution the time step Δt should be of the order of the dielectric relaxation time and the space steps Δx, Δy should not be significantly larger than a Debye length, both of which are functions of doping density and dielectric permittivity.

The linearised scheme described above is a popular choice and has been successfully applied in many simulations. However, the linearised discretization can lead to substancial errors in regions of high electric field and is particularly sensitive in highly doped devices. The source of the error can be traced to the assumption that the carrier density varies in a linear fashion between adjacent mesh nodes. In order to demonstrate this problem and determine a limit of validity for the linearised scheme, consider a one-dimensional Schottky barrier junction under zero bias conditions. In these circumstances there should be no current flow. This implies that the drift and diffusion components of the current density J_{n_x} should balance,

$$J_{n_x} = -qn\mu_n \frac{\partial \psi}{\partial x} + qD_n \frac{\partial n}{\partial x} = 0 \tag{4.27}$$

that is assuming uniform mesh spacings

$$n_{i+1/2}\mu_{n_{i+1/2}} \frac{\psi_{i+1} - \psi_i}{\Delta x} = D_{n_{i+1/2}} \frac{n_{i+1} - n_i}{\Delta x} \tag{4.28}$$

where $n_{i+1/2}$, $\mu_{n_{i+1/2}}$ and $D_{n_{i+1/2}}$ are determined from linear interpolation of the values at the adjacent nodes i and $i+1$. Substituting for $n_{i+1/2}$ and the diffusion coefficient $D_{n_{i+1/2}}$ using the Einstein relationship,

$$\psi_{i+1} - \psi_i = \frac{2kT}{q}\frac{n_{i+1} - n_i}{n_{i+1} + n_i} \tag{4.29}$$

where k is Boltzmann's constant. If it is assumed that the depletion region is abrupt with $n_{i+1} = N_D$ and $n_i \geq 0$ (the worst case),

$$\psi_{i+1} - \psi_i \leq \frac{2kT}{q} \tag{4.30}$$

Hence the potential difference between nodes cannot exceed $2kT/q$ Volts (about 52 mV at room temperature) or the linearised discretization will produce negative values of electron density n. This manifests itself as numerical instablity in the solution. The possibility of this situation arising can be avoided by assuming a different type of variation (non-linear) between the mesh points. Scharfetter and Gummel demonstrated that by allowing the electron density to follow an exponential variation between mesh points, errors due to the discretization could be avoided [8]. The current density equations for $\mathbf{J}_n$ and $\mathbf{J}_p$ are treated as differential equations in n and p and it is assumed that J_n, J_p, μ_n, μ_p and E are constant between mesh points. In one-dimension the electron current density may then be expressed as,

$$J_{n_{x_{i+1/2}}} = -\frac{qD_n}{a_i}\left\{n_{i+1}B\left[\frac{q}{kT}(\psi_{i+1} - \psi_i)\right] - n_iB\left[\frac{q}{kT}(\psi_i - \psi_{i+1})\right]\right\} \tag{4.31}$$

where a_i is the mesh spacing between $x=i$ and $x=i+1$ (Δx) and $B(x)$ is the Bernoulli function,

$$B(x) = \frac{x}{e^x - 1} \tag{4.32}$$

In order to speed up the computation process it is useful to evaluate the Bernoulli function using a set of predeterimined approximations for different ranges of x. This is necessary to avoid the possibility of underflows and overflows. It is assumed here that the Diffusion coefficient follows the Einstein relationship. It may be noted that if the potential difference between nodes is small then this equation approaches the standard difference equation form. The full expansion of the current continuity equation is lengthy and is given in Appendix 1. The apparent increase in execution time required by the evaluation of the Bernoulli functions is offset by the possibility of using fewer nodes and larger time-steps than in the linear discretised schemes. Hence, the Scharfetter-Gummel scheme is only slightly slower than linear discretized schemes and affords greater accuracy.

The schemes of Scharfetter and Gummel [8] and Slotboom [7] produce positive definite matrices and do not lead to coefficient matrices with complex eigenvalues and hence can be solved using SOR techniques (which often have a faster rate of convergence). Both approaches have been used extensively in finite-difference simulations.

4.3. Methods of Solving Finite-Difference Equations

Simultaneous equations formed by the finite-difference expressions for the Poisson and Çontinuity equations can be solved by direct or iterative techniques. A direct method of solving tridiagonal (and pentadiagonal) sets of simultaneous equations in which only the non-zero entries in the coefficient matrix need be stored, can be adapted from the Gaussian elimination technique. Another commonly used technique for the solution of the Poisson equations is based on the Fast Fourier Transform (FFT) [15] This method requires a moderately lengthy coding compared with iterative methods.

Large systems of simultaneous equations with sparse coefficient matrices are suited to iterative methods which are normally faster, require less memory and are more accurate than applying direct methods. Three common iterative methods are Gauss-Seidel, successive-over-relaxation (SOR) and the cyclic Chebyshev method.

The Gauss-Seidel method requires a diagonally dominant or positive-definite coefficient matrix to guarantee convergence. For example, in the case of the two-dimensional Poisson equation solved over a uniform mesh, the Gauss-Seidel method provides an improved estimate of $\psi_{i,j}$ in terms of the latest value at the surrounding four points,

$$\psi_{i,j} = \left[\frac{\psi_{i+1,j} + \psi_{i-1,j}}{\Delta x^2} + \frac{\psi_{i,j+1} + \psi_{i,j-1}}{\Delta y^2} - \frac{q}{\varepsilon_o \varepsilon_r}(N_{D_{i,j}} - n_{i,j})\right] \Big/ \left[2\left(\frac{1}{\Delta x^2} + \frac{1}{\Delta y^2}\right)\right] \tag{4.33}$$

The difference between the left-hand side and the right-hand side of equation (4.33) for the latest value of $\psi_{i,j}$ is called the residual. The residual is zero for the exact solution. For equal mesh increments in the x and y directions, the latest value of the potential $\psi^r_{i,j}$ is expressed in terms of the previous estimate $\psi^r_{i,j}$ and the residual $R^{r-1}_{i,j}$ as,

$$\psi^r_{i,j} = \psi^{r-1}_{i,j} + \frac{1}{4} R^{r-1}_{i,j} \tag{4.34}$$

where the superscript r denotes the iteration number.

Successive-over-relaxation can be used to increase the ratio of convergence of the Gauss-Seidel method by multiplying the residual by a relaxation factor ω,

$$\psi^r_{i,j} = \psi^{r-1}_{i,j} + \frac{\omega}{4} R^{r-1}_{i,j} \tag{4.35}$$

The relaxation factor ω is selected to provide either a minimum error in a specific number of iterations or to ensure convergence of the solution to a required tolerance in the miimum number of iterations. The latter criterion is normally used to select ω and it is found that for a rectangular region of $n \times m$ mesh points, with a diagonally dominant coefficient matrix, the optimum value of ω, denoted ω_{opt}, is related to the smaller root of the quadratic equation,

$$z^2\left(\cos\frac{\pi}{m} + \cos\frac{\pi}{n}\right)^2 - 4z + 1 = 0 \tag{4.36}$$

where

$$\omega_{opt} = 4z \tag{4.37}$$

The relaxation factor calculated for a diagonally dominant coefficient matrices lies in the range of 1 to 2 for SOR [13]. There is no corresponding analytical method for selecting the relaxation factor for non-rectangular regions. Under-relaxation ($0<\omega<1$) is usually used where there is no diagonal dominance in the coefficient matrix. It is useful to re-express equation (4.35) in the form,

$$\psi_{i,j}^{r} = \psi_{i,j}^{r-1} + \omega(\tilde{\psi}_{i,j}^{r} - \psi_{i,j}^{r-1}) \tag{4.38}$$

where $\tilde{\psi}_{i,j}^{r}$ is the value obtained from equation (4.33).

In the cyclic Chebyshev method, the value of the relaxation factor ω is changed every half-iteration to improve the rate of convergence. This relaxation factor is defined for a square $n \times n$ mesh as

$$\omega^{r+0.5} = \frac{1}{1 - 0.25\lambda^2\omega^{r}} \tag{4.39}$$

where

$$\lambda = \cos\frac{\pi}{n} \tag{4.40}$$

The initial rate of convergence is far higher for the Chebyshev method than for the straightforward SOR scheme. However, if few iterations are required to reach the specified convergence criteria then the extra calculation required for ω offsets the time saving obtained by this method.

An important advantage of iterative methods is that round-off errors in the solution are limited to those incurred in the final iteration and are normally negligible. This is because the values used in the last iteration are treated as initial estimates and updated independently of the rounding error. Direct elimination methods are subject to cumulative round off errors which can become significant for large coefficient matrices or for ill-conditioned systems of equations.

There are several solution techniques available for solving large systems of simulataneous equations typical of those generated by the finite-difference method. The most efficient solution technique is the one-step block SOR method, which is relatively easy to implement and requires half as much storage as the alternating direction implicit (ADI) scheme. LU decomposition methods have also been used successfully to solve the non-linear current continuity and Poisson equations.

The three principal approaches to solving the coupled Poisson and continuity equations are the simultaneous method, the alternating method (Gummel's method) and the sequential solution method. The choice of method is partly determined by whether a steady-state (dc) solution is required or alternatively a transient or ac time-dependent solution is sought. The simultaneous method produces a combined solution for both the Poisson and continuity equations simultaneously and utilizes a matrix twice as large as the individual matrices, requiring substancial computational effort. The Gummel algorithm [6] applied to steady-state solutions, which is in fact a modified version of Newtons method for fixed point iteration, starts by obtaining a converged solution for the Poisson equation to obtain the node potentials. The carrier densities are then updated using the new node potentials and the Poisson equation is solved once again. This procedure is repeated until the potential converges to within the a preset limit. In this way self-consistent solutions are obtained for the potential and carrier distribution. The sequential solution method, applied usually when a time-dependent solution is required, proceeds by first solving the Poisson equation and then solving the current continuity equation using the updated potentials. This process of sequentially solving the Poisson and current continuity equations repeats itself at each time step as the solution develops. There a many variations on this technique.

4.4. Boundary Conditions

The solution of the semiconductor equations requires suitable expressions for the boundary conditions. The potentials and carrier densities at contact nodes are usually determined using Dirichlet boundary conditions, where the variables are fixed at a predermined value. This type of boundary condition is easily implemented using finite-difference equations. For example for an ohmic contact on the surface $y = 0$ where $0 \leq x \leq 10$ the potential on the contact could be defined as

$$\psi_{i,0} = V_{applied} \qquad 0 \leq i \leq 10 \tag{4.41}$$

where $V_{applied}$ is the voltage applied to the contact as the boundary condition. The carrier concentrations may be defined in a similar manner.

The 'free' surfaces (boundaries other than contacts) are usually modelled using derivative boundary conditions, where for example it is assumed that there is no current flow across the boundary. This requires that the derivative of the electric field and carrier densities normal to the surface are zero. Derivative (Neumann) boundary conditions are generally more difficult to implement than Dirichlet boundary conditions and erroneous solutions may occur in the transition from between Neumann and Dirichlet boundaries.

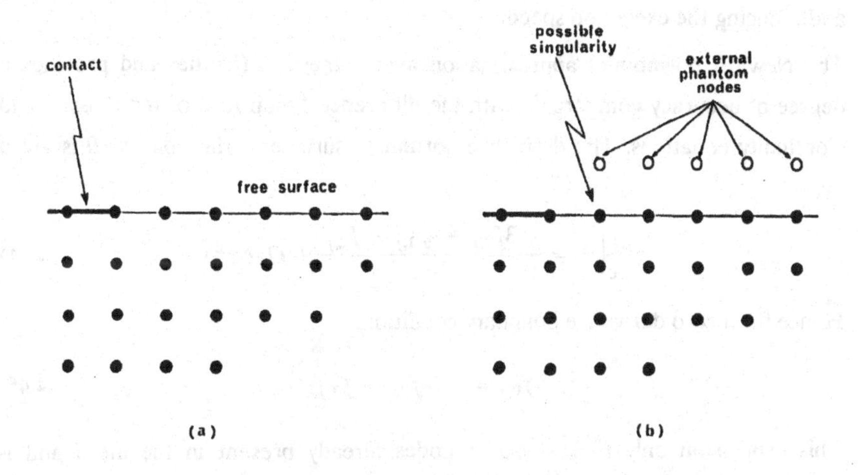

Figure 4.5 Boundary conditions for finite-difference meshes.
(a) Basic mesh in the region of a contact and free-surface
(b) Phantom nodes necessary for boundary conditions using Stirling polynomial approximations.

There are a number of ways of formulating derivative boundary conditions in finite-difference notation. Suitable boundary condition approximations are obtained from Newton polynomials and Stirling polynomials [16,17]. The approximation obtained from Stirling polynomials requires extra 'phantom' nodes outside the surface of the device (Figure 4.5) and leads to singularities in the solution near contacts. The Stirling derivative approximation for a surface parallel to the y axis at $x = 0$ is,

$$f'_{0,j} = \frac{f_{1,j} - f_{-1,j}}{2\Delta x} + O(\Delta x^2) \tag{4.42}$$

where the row of phantom nodes is situated parallel to the y axis at $x = -1$. The final term on the right-hand side of the equation indicates that the error is of order Δx^2. The derivative is usually set to zero for the potential ($E_{normal} = 0$) and carrier distributions and hence for the surface parallel to the y axis at $x = 0$,

$$f_{-1,j} = f_{1,j} \tag{4.43}$$

A further disadvantage of this type of boundary condition is that the size of the mesh is increased by 2 rows in each dimension, increasing storage requirements and reducing the execution speed.

The Newton polynomial approximation avoids these difficulties and provides a degree of accuracy comparable with the difference expansions of the Poisson and Continuity equations. The derivative normal to surface at the point x=0 is given by,

$$\left.\frac{df}{dx}\right|_{0,j} = \frac{-3f_{0,j} + 4f_{1,j} - f_{2,j}}{2\Delta x} + O(\Delta x^2) \tag{4.44}$$

Hence for a zero derivative boundary condition,

$$f_{0,j} = \frac{1}{3}(4f_{2,j} - f_{3,j}) \tag{4.45}$$

This expression only makes use of nodes already present in the mesh and is computationally more efficient than the Stirling approximation. The use of this type of boundary conditions means that the coefficient matrix does not satisfy the diagonal dominance conditions necessary to guarantee convergence in interative solutions. However despite this irregularity, convergence is normally possible although the optimum relaxation factor must be determined empirically. Expressions (4.42) and (4.44) can be modified and applied to any current free surface.

Simulations based on electron temperature or energy transport models will require derivative boundary conditions at the current free surfaces for the electron temperature or energy respectively. Dirichlet boundary conditions may be used for the electron energy/temperature at the contacts.

Contacts may also be modelled using more detailed relationships. In particular, Schottky contacts may be modelled using a combination of Dirichlet and Neumann boundary conditions derived from diffusion, thermionic emission or thermionic-emission-diffusion models [18,19]. The importance of modelling accurately the contacts (or other boundaries) cannot be overstated as the solution obtained from the simulation will be determined by the boundary conditions used. This apparently obvious statement is often forgotten when over-simplified contact and surface models are used as boundary conditions in practical device structures. For example, a forward biased Schottky barrier interfaced with a highly doped active layer is not represented accurately by the diffusion model. It is also erroneous to neglect the effects of surface depletion in many planar devices, even though the majority of researchers frequently do. It is also questionable whether the simple ohmic contact model described previously is adequate for all situations. The boundary conditions chosen for a particular device model should be carefully chosen to reflect the contact and surface properties of the device.

4.5. Examples of Finite-Difference Simulations

The finite-difference technique is a popular choice for the numerical soltion of the semiconductor equations because of the simplicity of implementing the numerical schemes. The majority of early numerical device simulations were carried out using this method and it is still one of the most common techniques used today. There are many published examples of device models which use this approach for one-, two- and three-dimensional simulations.

One-dimensional simulations are adequate for device models for many conventional semiconductor devices, where the current flow and electric field a predominantly unidirectional. This applies particularly to vertical device structures where the active region is sandwiched between relatively large contacts, Figure 4.6(a). In contrast, planar and surface orientated devices with contacts on the same surface, Figure 4.6(b), tend to have at least two-dimensional field and carrier distributions and are not generally well suited to one-dimensional models (there are exceptions to this case, when very long devices are considered). Two terminal vertical devices have traditionally been modelled using one-dimensional simulations. Devices in this category include p-n junction diodes [20], Schottky varactor diodes [19] transferred electron devices (TEDs) [21,22,23,24,25,26] and

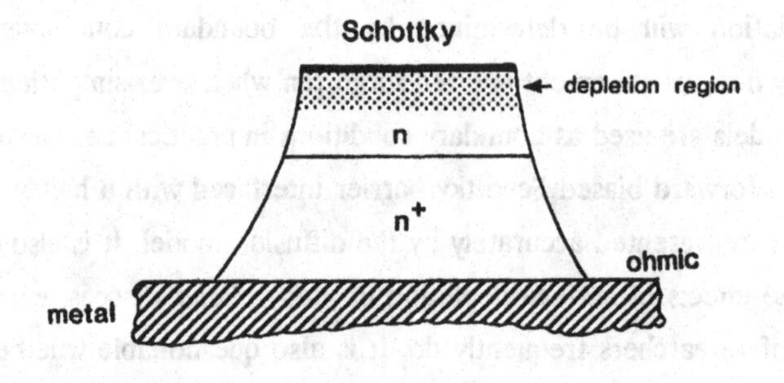

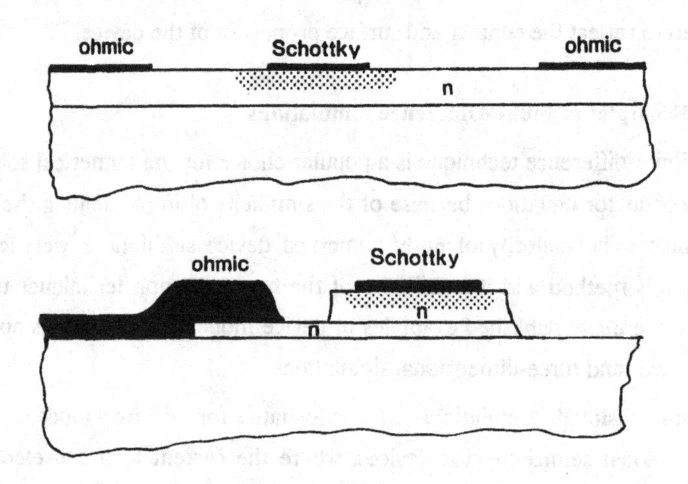

Figure 4.6 (a) vertical device structure (b) planar and non-planar surface orientated structures

IMPATT's [27,28].

The transferred electron device, which is fabricated from GaAs or InP, is used in a wide variety of microwave and millimetre wave subsystems as the active device for oscillators and reflection amplifiers. The key feature of the TED is that when it is biased correctly in a microwave cavity it is capable of producing current oscillations at microwave frequencies (usually in the range 10 GHz to 60 GHz). This device has been extensively investigated using numerical simulations which provide a dramatic picture of the physical phenomena involved in the operation of the device. TED's, also know as Gunn diodes, are usually fabricated from epitaxial layers of GaAs in an $n^+ - n - n^+$ layer structure, Figure 4.7, and are majority carrier devices. More recently there has been interest in using InP TED's to generate millimetre wave power at frequencies above 90 GHz. A full description of the operation and characteristics of transferred electron devices is available in other texts [29,30]. The operation of TED's is characterised by the presence of a charge domain, consisting of a non-uniformity in the electron and carrier distributions. In some circumstances the domain will travel through the device from the anode towards the cathode, although this is not always the case. TED's can operate in a variety of modes, determined by the conditions governing the formation and extinction of the domains. One-dimensional models are used to represent this structure because domains propagate between the parallel ohmic contacts and because the length of the active n layer is usually much smaller than the device diameter.

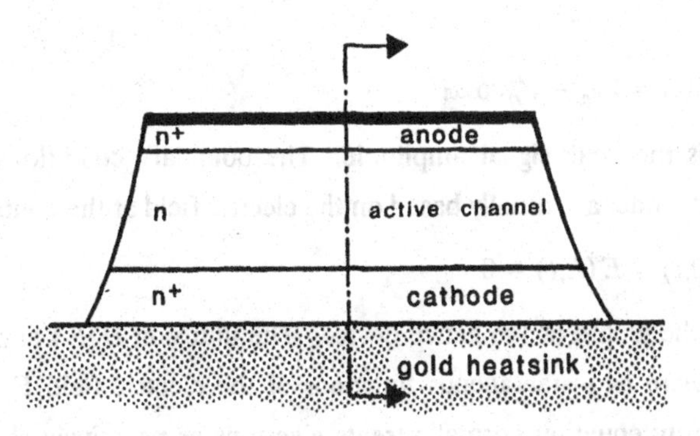

Figure 4.7 Transferred electron device structure (vertical device)

The TED is an example of a device whose dynamic operation can only be satisfactorily theoretically studied using numerical simulations. Since the TED operates in a large-signal mode, equivalent circuit models consisting of time-averaged elements, can only be obtained from measurements or simulations and are of limited value. Time-domain numerical simulations of TED's can be used to predict operating characteristics such as microwave output power, noise performance and stability, efficiency and tuning characteristics. TED simulations often simplified by eliminating the electron density from the current density expression by combining Poisson's equation with the current density expression and neglecting generation and recombination effects [26,30]. An example of a one-dimensional model based on this arrangment is described by the current density equation

$$J(t) = qv(E)\left[\frac{\varepsilon}{q}\frac{\partial E(x,t)}{\partial t} + N_D(x)\right] + qD(E)\left[\frac{\varepsilon}{q}\frac{\partial^2 E(x,t)}{\partial x^2} + \frac{\partial N_D(x)}{\partial x}\right] + \varepsilon\frac{\partial E(x,t)}{\partial t} \quad (4.46)$$

For this one-dimensional model the electric field E is determined by the applied voltage $V(t)$,

$$V(t) = \int_0^L E(x,t)\,dx \quad (4.47)$$

where

$$V(t) = V_{dc} - V_{rf}\cos\omega t \quad (4.48)$$

where V_{rf} is the peak signal amplitude. The boundary conditions for this one-dimensional model are usually based on the electric field at the contacts, so that,

$$E(0,t) = E(L,t) = 0 \quad (4.49)$$

where L is the length of the active device. The diffusion $D(E)$ and velocity $v(E)$ are shown here as instantaneous functions of the electric field E_x. The form of current density equation normally treats electrons as negatively charged particles so that $\mathbf{v} = -\mu\mathbf{E}$. It is often more convenient to treat electrons as positively charged particles in unipolar simulations, which requires the sign of the diffusion term to be changed in the current density equation.

The time-domain simulation of TED's can be used to provide information on both the device and device-circuit interaction. In addition to the carrier, field, current density and potential profiles in the device, it is relatively easy to extract information on the microwave performance of the device embedded in a simple circuit. This can be achieved in two ways. Firstly, by stimulating the device with a sinusoidal voltage as above, the corresponding current waveform can be obtained and Fourier analysed. This information can be used to obtain an equivalent device admittance ($I(\omega)/V(\omega)$) at specific frequencies ω and signal amplitudes A. The device admittance $-\overline{G}(A,\omega) + \overline{B}(A,\omega)$ can then be plotted over regions of negative conductance (where microwave power is being generated) as a function of frequency and signal amplitude to yield a device surface as in Figure 4.8. The device surface can be used in conjunction with the equations which describe steady state oscillation conditions for two-terminal negative conductance oscillators to predict the operating charecteristics of a transferred electron oscillator,

$$-\overline{G}(A,\omega) + G(\omega) = 0 \tag{4.50}$$

$$\overline{B}(A,\omega) + B(\omega) = 0 \tag{4.51}$$

where $G(\omega) + B(\omega)$ is the admittance of the microwave load. Alternatively, specific operating points and oscillator behaviour can be examined by connecting a time-domain circuit model to the device model, utilising the circuit voltages as the instantaneous bias applied to the device. This approach has been used to characterise prototype oscillator circuits [32]. A more detailed description of both these approaches is described in [32,33].

Recent interest has centred on modelling millimetre wave transferred electron oscillators operating in fundamental and harmonic modes. This requires a more detailed model, incorporating hot electron effects. The semiconductor equations in this type of model include energy and momentum relaxtion effects described in Chapter 6.

There has been considerable simulation effort devoted to GaAs TEDs over the last eighteen years. It has been established that the nature of the contact properties, doping, mobility, and interfacial fields, particularly at the cathode, determine the operational behaviour of the device. Doping notches near the cathode are frequently incorporated into the models to approximate the domain nucleating mechanism. Typical results obtained from time-dependent one-dimensional finite-difference numerical simulations of GaAs TED's are shown in Figure 4.9

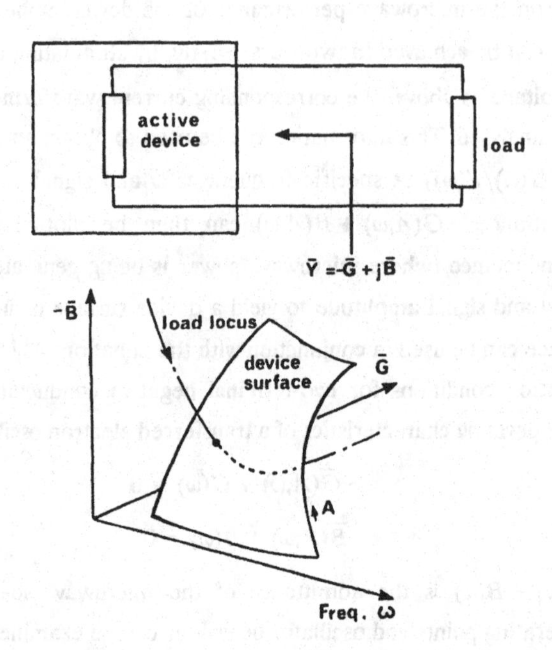

Figure 4.8 One-port oscillator model and associated device surface used to characterise microwave oscillators

and Figure 4.10. Models have been developed which take into account the effects of space and time-dependent ionized impurities due to carrier generation-recombination mechanisms [34]. Indium phosphide TED's have also been modelled to investigate the higher efficiences of this device compared with GaAs TED's [35]. The increasing interest in using InP devices at millimetre frequencies makes the use of numerical simulations even more important for the successful development of this type of oscillator.

Simulations based on finite-difference schemes have been used to investigate a wide variety of surface-orientated devices including BJTs [4,35], JFETs [36], MOSFETs [37,38,39,40,41], MESFETs [42,43,44], planar TEDs [45,46] and thyristors [47]. The MOSFET (also known as the IGFET) is the most prolifically studied three terminal device because of its crucial role in the VLSI technology. Small geometry MOSFETS are used to obtain high packing densities and high speed operation in integrated circuits. At least two-dimensional models are

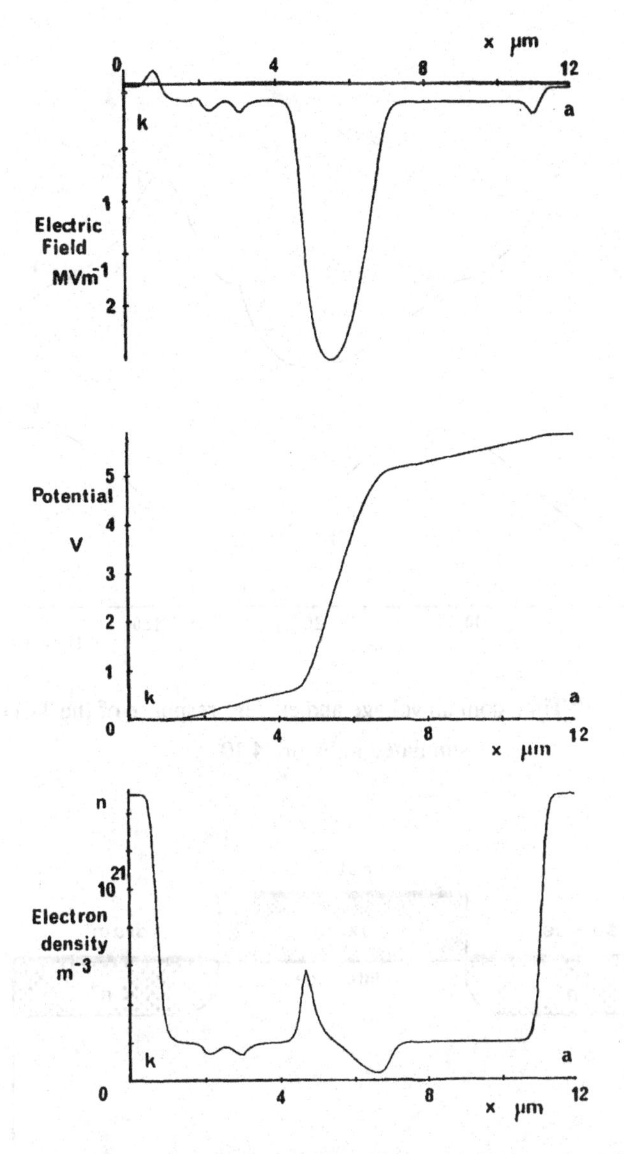

Figure 4.9 Typical results obtained from simulation of a microwave transferred electron device (X Band)

required for these small-geometry MOSFETs to account for the multi-dimensional nature of the field and carrier distributions.

MOSFET operation involves both electron and hole transport, although hole current is negligible in most cases. A two-dimensional model of a silicon n-

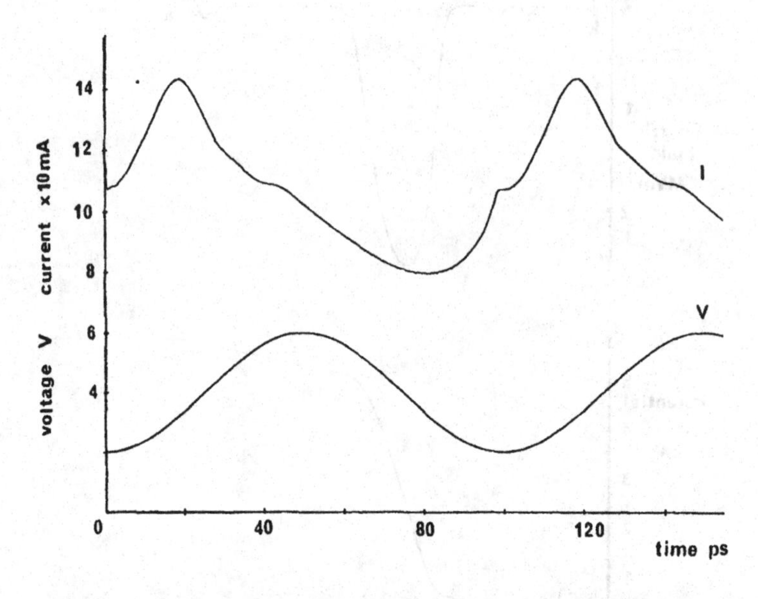

Figure 4.10 Time domain voltage and current response of the TED simulated in Figure 4.10

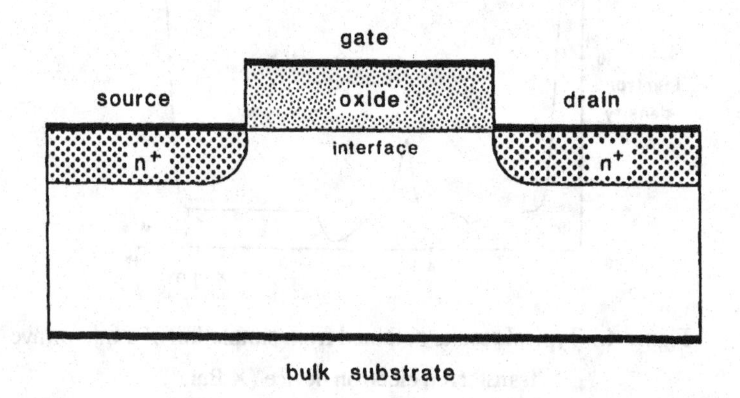

Figure 4.11 Two-dimensional silicon n-channel MOSFET model

channel MOSFET is shown in Figure 4.11, showing heavily doped n^+ contact regions, the p-type bulk and the semiconductor-oxide interface beneath the gate. In the model the electron densities at the source and drain contacts are set equal to the doping concentrations. The potential on the source, drain and bulk contacts is determined by the applied bias voltage plus the built-in potential, which assumes these are ideal ohmic contacts. The gate potential is set to the applied bias voltage minus the flatband voltage. The boundary condition for the oxide region assumes that no space-charge exists in this region and hence only the Laplace equation has to be solved,

$$\nabla^2 \psi = 0 \tag{4.52}$$

Many authors assume a one-dimensional electric field in the oxide (for example [48]), although this approximation is not valid for small devices [41]. At the interface the following boundary condition is applied following Gauss' law,

$$\varepsilon_{oxide} \left. \frac{\partial \psi}{\partial y} \right|_{oxide} = \varepsilon_{s/c} \left. \frac{\partial \psi}{\partial y} \right|_{semiconductor} \tag{4.53}$$

The current-free vertical surfaces either side of the gate contact and at each end of the device model are assumed to have zero potential derivative boundary conditions normal to the surfaces. Surface states have also been included in some models [49], although their effect on the solution is very small and can be accounted for within the flatband voltage.

The numerical simulation of MOSFETs is well established and an increasing number of simulation software packages are becoming available. The MOSFET is possibly the first device to merit the use of sophisticated numerical simulations in industry as part of the design process, as well as in a research environment. Examples of two-dimensional application packages for MOSFET modelling include CADDET [50], GEMINI [51] and MINIMOS [52]. MINIMOS is capable of analysing coupled devices in monolithic silicon circuits such as NMOS and CMOS inverters. This represents an important development in the use of physical modelling for small-scale circuit design as well as discrete device characterisation.

MINIMOS, a two-dimensional MOSFET simulator, is one of the most popular simulation packages developed in recent years, with over 300 institutions now using the package. Hot electron transport (discussed in chapter 6) has been recently added to this simulation. The program utilises an adaptive automatic mesh refinement algorithm. The semiconductor equations are scaled using the singular perturbation approach introduced by Markowich [53]. Standard finite-difference discretisation is utilised. The three coupled continuity and Poisson equations are solved using an algorithm similar to Gummel's iterative method. The linearised Poisson equation is solved using the block cyclic Jacobi conjugate gradient method (BCJCG) [54]. The linearised continuity equations are solved using Gaussian elimination with a checkerboard ordering algorithm. Stone's method for solving all three equations was used in earlier versions, but poor convergence was experienced under some circumstances which has been overcome using the BCJCG method. A further modification to the latest version of MINIMOS is the use of Mock's recent algorithm for calculating terminal currents, which minimises the sensitivity due to mesh spacing [55]. Typical results for a device with a uniformly doped substrate ($10^{21} m^{-3}$) are shown in Figure 4.12. The n^+ implants below the contacts are doped at $5 \times 10^{25} m^{-3}$, with the p-n junction approximately 0.3 μm below the surface [52].

Three-dimensional effects have been shown to significantly influence the operation of very small geometry VLSI MOSFETs (with gate widths of 10μm or less and gate lengths under 1μm), where the channel width of of the same order as the depletion layer thickness. Furthermore, for this scale of integration the device structures appear highly non-planar. Husain and Chamberlain [56] have described a three-dimensional finite-difference simulation called WATMOS, for modelling small geometry MOSFETs. It has been demonstrated that three-dimensional fringing fields effect the threshold potential of small geometry devices and that the effective width of the device is non-uniform throughout the channel region.

Metal semiconductor field effect transistors (MESFET's) have been extensively investigated using two-dimensional finite-difference numerical simulations. The GaAs MESFET forms the foundation of active microwave monolithic integrated circuits (MMIC's) and has largely replaced silicon bipolar transistors and junction FETs (JFETs) as a microwave transistor above 4 GHz. GaAs MESFET technology (along side the developing HEMT technology) forms the basis of

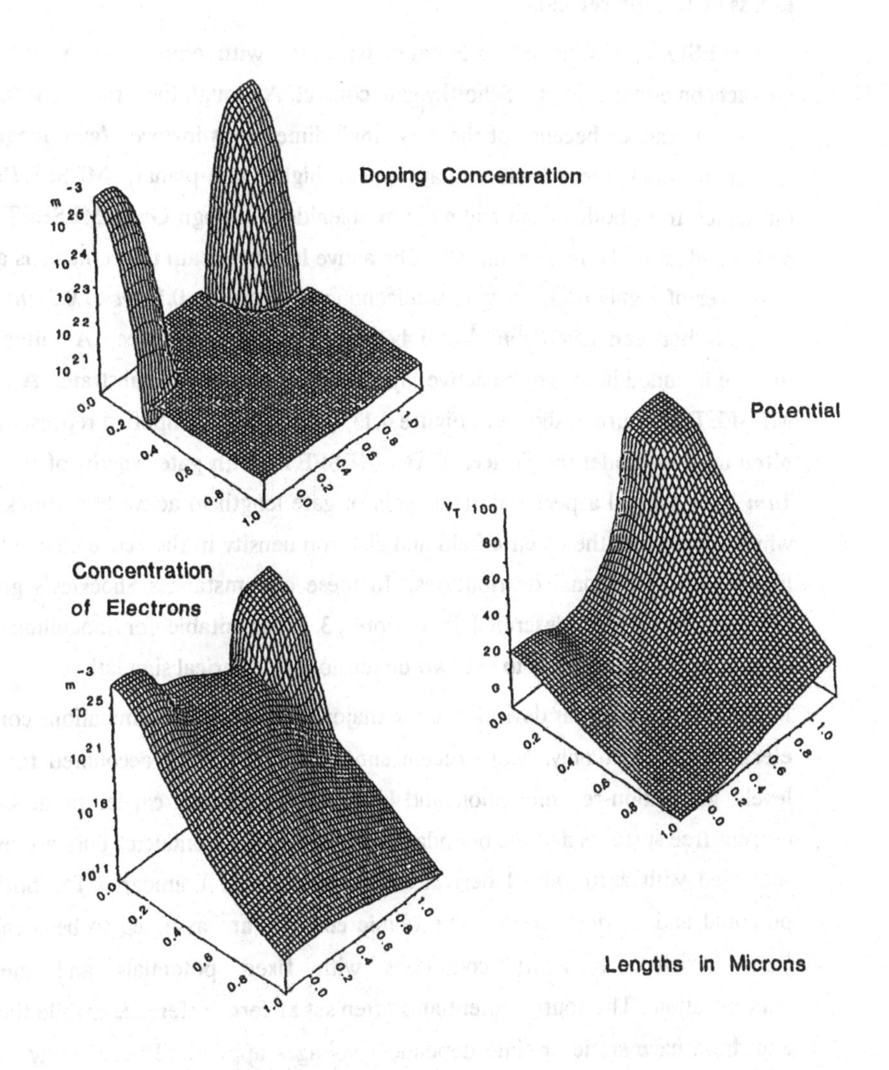

Figure 4.12 Typical MOSFET simulation results

gigabit rate integrated circuits for future digital applications. Medium scale integrated circuits have already been demonstrated to operate at frequencies in excess of 10 Gbit/second.

The MESFET is a surface orientated transistor with ohmic source and drain contacts on either side of a Schottky gate contact. Although the structure is termed planar, in practice because of the very small dimensions involved (gate lengths of less than $1\mu m$), the structure may appear highly non-planar. MESFET's are fabricated from both silicon and gallium arsenide, although GaAs MESFET's are by far and away the most common. The active layer beneath the contacts is a very thin layer of highly doped n type semiconductor, typically $0.1\ \mu m$ to $0.2\ \mu m$ thick doped at between $1.5\times10^{23} m^{-3}$ and $4\times10^{23} m^{-3}$ for GaAs devices. A buffer layer may be included between the active layer and semi-insulating substrate. A typical MESFET structure is shown in Figure 4.13, along with a simplified representation often used to model the device. GaAs MESFET's with gate lengths of less than $1\mu m$ have a small aspect ratio (the ratio of gate length to active layer thickness), which means that the electric field and electron density in the active channel have highly two-dimensional distributions. In these circumstances Shockley's gradual channel model [57], described in Chapter 3, is unsuitable for modelling these devices and it is necessary to use two-dimensional numerical simulations.

MESFETs are unipolar devices and the majority of MESFET simulations consider electron transport only. More recent simulations [58] have accounted for deep levels, generation-recombination and breakdown with two carrier models. The current-free surfaces and the boundaries inside the semiconductor bulk are usually modelled with zero valued derivatives normal to the boundaries for both the potential and carrier density. The ohmic contacts are assumed to be ideal and have Dirichlet boundary conditions with fixed potentials and electron concentration. The source potential is often set at zero (reference), while the gate and drain have static or time dependent voltages applied. The Schottky barrier gate is characterised by the difference in work function between the metal contact and the semiconductor material, with a potential ψ_G fixed at the applied voltage minus the built in voltage ϕ_{Bi}, and electron concentration n_G often based on a

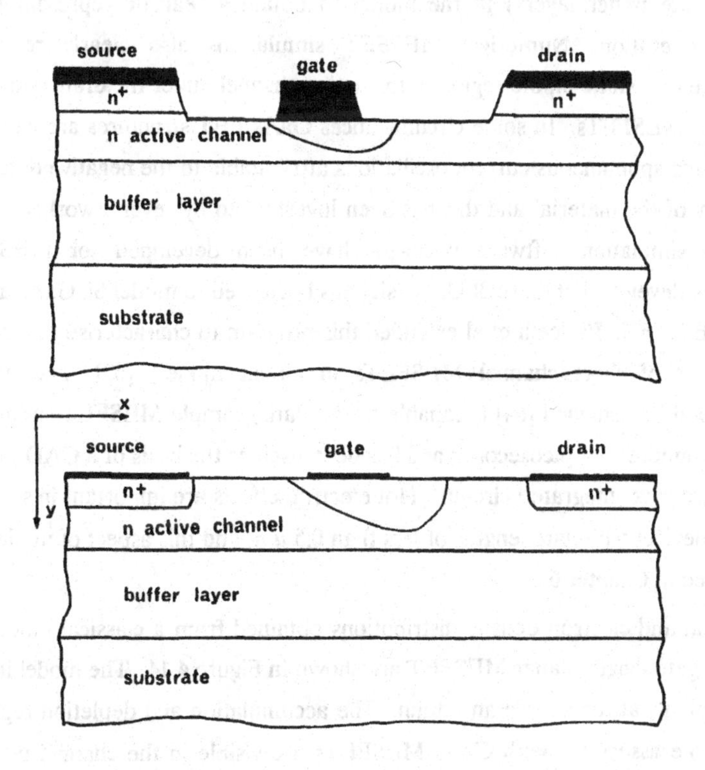

Figure 4.13 Typical MESFET structure and simplified representation

diffusion model,

$$\psi_G = v_G(t) - \phi_{Bi} \tag{4.54}$$

$$n_G = N_D \exp\left(-q\phi_{Bi}/kT\right) \tag{4.55}$$

More detailed Schottky barrier boundary conditions have been modelled using thermionic emission-diffusion theory and thermionic emission theory [18,19,59].

Numerical simulations have been used to provide insight into the operation of short gate-length MESFET's and investigate the influence of different device structures on electrical performance. The effects of different doping profiles, buffer layers and semi-insulating substrates have been considered (see for example

[3,43,58,60]. It is apparent that it is essential to include substrates (and where appropriate buffer layers) in the model to obtain a realistic representation of device operation. Numerical MESFET simulations also clearly reveal the presence of a static dipole region in the active channel under the drain end of gate in GaAs MESFETs. In some circumstances GaAs FET structures are capable of supporting spontaneous current oscillations attributable to the negative differential mobility of the material and this has been investigated by several workers [42,61]. Several simulation software packages have been developed for MESFET's. CUPID, developed at Cornell University has been used to model Si, GaAs and InP MESFETs [62]. Faricelli et al extended this program to characterise the transient behaviour of short channel MESFETs in circuit models [63]. The program developed by Snowden [64] is capable of simulating simple MESFET circuits over several hundreds of picoseconds and has been used as the basis of a CAD program for microwave integrated circuits. Hot electron effects are important in short gate length devices with gate lengths of less than 0.5 μm, and this aspect of modelling is discussed in Chapter 6.

Potential and electron density distributions obtained from a classical model for a 0.5 μm gate length planar MESFET are shown in Figure 4.14. The model includes n^+ implants at the source and drain. The accumulation and depletion regions of the dipole associated with GaAs MESFETs are visible in the channel under the drain end of the gate.

The drive towards smaller device geometries and optimization of existing structures has led to a proliferation of surface orientated non-planar structures (for example recessed gate MESFET's, short gate length MOSFET's). Non-uniform meshes are used in conjunction with finite-difference schemes to model non-planar device structures [9,60,65,66,67,68]. The descretization of Poissons equation for non-uniform finite-difference meshes is achieved by integrating the over the local area surrounding each node and then approximating the integrals to produce a 'five-point' difference approximation similar to that discussed previously, but with unequal mesh spacings. The discretized Poisson equation may then be solved using the highly efficient one-step block SOR technique. This technique has in particular been applied to the study of non-planar MOS

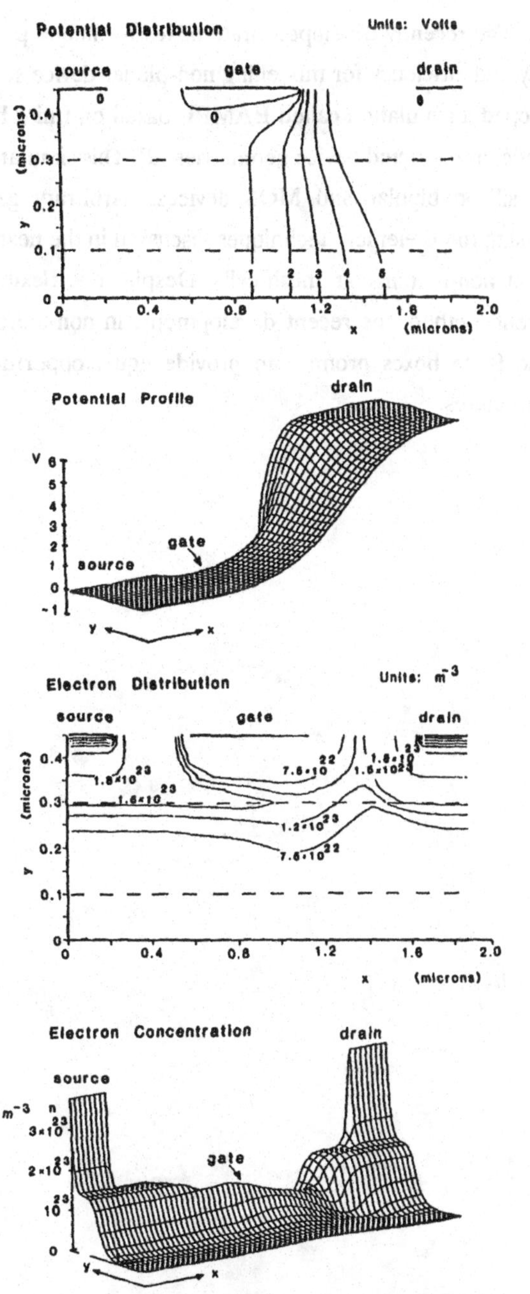

Figure 4.14 Simulation results obtained from a drift-diffusion model of a 0.5 μm gate length MESFET

transistors. The recently developed finite boxes technique provides a high degree of flexibility and efficiency for modelling non-planar device structures. Franz et al have developed a simulation called BAMBI, based on finite boxes, which may be used to model unrestricted device geometries [9]. This simulation has been applied applied to silicon bipolar and MOS devices. Arbitrary geometries are often modelled using finite-element techniques discussed in the next chapter, which take advantage of non-rectangular mesh-cells. Despite this flexibility inherent in the finite-element method, the recent developments in non-uniform finite-difference meshes and finite boxes promise to provide equal opportunities for modelling arbitrary structures.

References

[1] De Mey, G., "Determination of the electric field in a Hall generator under influence of an alternating magnetic field", Solid State Electron., 17, pp.977-979, 1974

[2] Oh, S-Y. and Dutton, R.W., "A simplified two-dimensional numerical analysis of MOS devices - DC case", IEEE Trans. Electron Devices, ED-27, pp.2101-2108, 1980

[3] Reiser, M., "Large-scale numerical simulation in semiconductor device modelling", Comp. Meth. Appl. Mech. Engng, 1 , pp.17-38, 1972

[4] Slotboom, J.W., "Computer-aided two-dimensional analysis of bipolar transistors", IEEE Trans. Electron Devices, ED-20. 669-679, 1973

[5] Richtmeyer, R.D. and Marton, K.W., Difference Methods for Initial Value Problems, New York: Wiley-Interscience, 1957

[6] Gummel, H.K., "A self-consistent iterative scheme for one-dimensional steady state transistor calculations", IEEE Trans. Electron Devices, ED-11, pp.455-465, 1964

[7] Slotboom, J.W., "Iterative scheme for 1- and 2- dimensional DC transistor simulation", Electron. Lett., 5, 677-678, 1969

[8] Scharfetter, D.L. and Gummel, H.K., "Large-signal analysis of a silicon Read diode oscillator", IEEE Trans. Electron Devices, ED-16, pp.64-77, 1969

[9] Franz, A.F., Franz, G.A., Selberherr, S., Ringhofer, C. and Markowich, P., "Finite Boxes - a generalization of the finite-difference method suitable for semiconductor device simulation", IEEE Trans. Electron Devices, ED-30, pp.1072-1082, 1983

[10] Reiser, M., "A two-dimensional numerical FET model for DC, AC, and large-signal analysis", IEEE Trans. Electron Devices, ED-20, pp.35-45, 1973

[11] Dubock, P., "D.C. numerical model for arbitrarily biased bipolar transistors in two dimensions", Electron. Lett., 6, pp.53-55, 1970

[12] Snowden, C.M., Doades, G.P., Howes, M.J. and Morgan, D.V., "Two terminal characterization of microwave monolithic elements", Proc. 11th European Microwave Conference, pp.633-638, 1981

[13] Varga, R.S., Matrix Iterative Analysis, Englewood Cliffs, NJ: Prentice-Hall, 1962

[14] Reiser, M., "A two-dimensional numerical FET model for dc- ac- and large-signal analysis", IBM Research J., RZ 499, April 1972

[15] Hockney, R.W., "The potential calculation and some applications", Meth. Comput. Phys., 9, pp.135-211, 1970

[16] Gerald, C.F., Applied Numerical Analysis, New York: Addison-Wesley, 1973

[17] Fenner, R.R., Computing for Engineers, London: Macmillan, 1978

[18] Snowden, C.M., Howes, M.J. and Morgan, D.V., "Large-signal modeling of GaAs MESFET operation", IEEE Trans. Electron Devices, ED-30, pp.1817-1824, 1983

[19] Tasker, P.J., "Ion implanted GaAs hyperabrupt varactor diodes", Ph.D. Thesis, University of Leeds, 1983

[20] De Mari, A., "An accurate numerical steady state one-dimensional solution of the pn junction", Solid State Electron, 11, pp.33-58, 1968

[21] McCumber, D.E. and Chynoweth, A.G., "Theory of negative-conductance amplification and of Gunn instabilities in "two-valley" semiconductors", IEEE Trans. Electron Devices, ED-13, pp.4-21, 1966

[22] Kroemer, H., "Nonlinear space charge domain dynamics in a semiconductor with negative differential mobility", IEEE Trans. Electron Devices, ED-13, pp.27-40, 1966

[23] Thim, H.W., "Computer study of bulk GaAs devices with random one-dimensional doping fluctuations", J.Appl. Phys., 39, pp.3897-3905, 1968

[24] Robrock, R.B., "Analysis and simulation of domain propagation in non uniformly doped bulk GaAs", IEEE Trans. Electron Devices, ED-16, pp.647-653, 1969

[25] Freeman, K.R. and Hobson, G.S., "The V_{fT} relation of CW Gunn effect devices", IEEE Trans. Electron Devices, ED-19, pp.62-70, 1972

[26] Lakshminarayana, M.R. and Partian, L.D., "Numerical simulation and measurement of Gunn device microwave characteristics", IEEE Trans. Electron Devices, ED-27, pp.546-552, 1980

[27] Mains, R.K., Haddad, G.I. and Blakey, P.A., "Simulation of GaAs IMPATT diodes including energy and velocity transport equations", IEEE Trans. Electron Devices, ED-30, pp.1327-1328, October 1983

[28] Scanlon, S.O. and Brazil, T.J., "Large-signal computer simulation of IMPATT diodes", IEEE Trans. Electron Devices, ED-28, No.1, pp.12-21, January 1981

[29] Hobson, G.S., The Gunn Effect, Oxford: Clarendon, 1974

[30] Sze, S.M., Physics of Semiconductor Devices, Wiley-Interscience, 1969

[31] Grubin, H.L., "Large-signal computer simulations of the contact, circuit and bias dependence of X-band transferred electron oscillators", IEEE Trans. Electron Devices, ED-25, pp.511-519, 1978

[32] Snowden, C.M., "Computer-aided design and characterisation of MMIC oscillators using physical device models", Proc. 14th European Microwave Conference (Publ. Sevenoaks: Microwave Exhibitions and Publishers Ltd.), pp.717-722, 1984

[33] Snowden, C.M., "Microwave F.E.T. oscillator development based on large-signal characterisation", PhD Thesis, University of Leeds, 1982

[34] Maksym, P.A. and Hearn, C.J., "Carrier generation-recombination in the Gunn effect", Solid State Electron., 21, pp.1531-, 1978

[35] Heimeier, H.H., "A two-dimensional numerical analysis of a silicon bipolar transistor", IEEE Trans. Electron Devices, ED-20, pp.708-714, 1973

[36] Kennedy, D.P. and O'Brien, R.R., "Computer-aided two-dimensional analysis of the junction field-effect transistor", IBM J.Res. Dev., 14, pp.95-116, 1970

[37] Oka, H., Nishiuchi, K., Nakamura, T. and Ishikawa, H., "Two-dimensional numerical analysis of normally-off type buried channel MOSFET's", IEEE Trans. Electron Devices, ED-27, pp.1514-1520, 1980

[38] Liu, S., Hoefflinger, B. and Pederson, D.O., "Interactive two-dimensional design of barrier-controlled MOS transistors", IEEE Trans. Electron Devices, ED-27, pp.1550-1558, 1980

[39] Mock, M.S., "A two-dimensional mathematical model of the insulated-gate field-effect transistor", Solid State Electron., 16, pp.601-609, 1973

[40] Yu, S.P. and Tantraporn, W., in VLSI Electronics Microstructure Science, Vol.3, ed. N.G. Einspruch, New York: Academic, pp.163-195, 1982

[41] Selberherr, S., Schutz, A. and Potzl, H.W., in Process and Device Simulation for MOS-VLSI Circuits, ed P. Antognetti et al, Amsterdam: Nijhoff, pp.490-581, 1983

[42] Yamaguchi, K., Asai, S. and Kodera, H., "Two-dimensional numerical analysis of stability criteria of GaAs FET's", IEEE Trans. Electron Devices, ED-23, pp.1283-1290, 1976

[43] Snowden, C.M., "Numerical simulation of microwave GaAs MESFETs", Proc. Int. Conf. on Simulation of Semiconductor Devices and Processes, Swansea: Pineridge, pp.405-425, 1984

[44] Snowden, C.M., "Two-dimensional modelling of non-stationary effects in short gate-length MESFETs", Proc. 2nd Int. Conf. on Simulation of Semiconductor Devices and Processes, Swansea: Pineridge, pp.544-558, 1986

[45] Goto, G., Nakamura, T. and Isobe, T., "Two-dimensional domain dynamics in a planar Schottky gate Gunn effect device", IEEE Trans. Electron Devices, ED-22, pp.120-126, 1975

[46] Doades, G.P., Howes, M.J., and Morgan, D.V., "Performance of GaAs device simulations at microwave frequencies", Proc. Int. Conf. on Simulation of Semiconductor Devices and Processes, Swansea: Pineridge, pp.353-367, 1984

[47] Berz, F., Gough, P.A. and Slatter, J.A.G., "The application of the enthalpy method to modelling the transient recovery of P-I-N diodes and G-T-O thyristors", Proc. Int. Conf. on Simulation of Semiconductor Devices and Processes, Swansea: Pineridge, pp.182-203, 1984

[48] Toyabe, T. and Asai, S., "Analytical models of threshold voltage and breakdown voltage of short channel MOSFET's derived from two-dimensional analysis", IEEE Trans. Electron Devices, ED-26, pp. 453-461, 1980

[49] Sutherland, A.D., "An algorithm for treating interface surface charge in the two-dimensional discretization of Poisson's equation for the numerical analysis of semiconductor devices such as MOSFET's", Solid State Electron., 23, 1085-1087, 1980

[50] Toyabe, T., Yamaguchi, K., Asai, S.and Mock, M., "A numerical model of avalanche breakdown in MOSFET's", IEEE Trans. Electron Devices, ED-25, pp.825-832, 1978

[51] Greenfield, J.A. and Dutton, R.W., "Nonplanar VLSI device analysis using the solution of Poisson's equation", IEEE Trans. Electron Devices, ED-27, pp.1520-1532, 1980

[52] Selberherr, S., Schutz, A. and Potzl, H.W., "MINIMOS - a two-dimensional MOS transistor analyzer", IEEE Trans. Electron Devices, ED-27, pp.1540-1550, 1980

[53] Markowich, P.A., The Stationary Semiconductor Device Equations, Wien-New York: Springer, 1985

[54] Hageman, L.A., Franklin, T.L. and Young, D.M., "On the equivalence of certain iterative acceleration methods", SIAM J.Numer.Anal., Vol. 17, pp.852-873, 1980

[55] Mock, M.S., Analysis of Mathematical Models of Semiconductor Devices, Dublin: Boole Press, 1983 10 in selbs 86 pap

[56] Husain, A. and Chamberlain, S.G., "Three-dimensional simulation of VLSI MOSFET's: the three-dimensional simulation program WATMOS", IEEE Trans. Electron Devices, ED-29, pp.631-638, 1982

[57] Shockley, W., "A unipolar 'field-effect' transistor", Proc. IRE, 40, pp.1365-1377, 1952

[58] Mottet, S. and Viallet, J.E., "Simulation of III-V devices on semi-insulating materials", Proc. 2nd Int. Conf. on Simulation of Semiconductor Devices and Processes, Swansea: Pineridge, pp.494-507, 1986

[59] Sangiorgi, E., Rafferty, C., Pinto, M. and Dutton, R., "", Proc. Int. Conf. on Simulation of Semicond Devices and Processes, Swansea: Pineridge, pp. 164-171, 1984

[60] Barton, T.M., Snowden, C.M. and Richardson, J.R., "Modelling of recessed-gate MESFET structures", Proc. 2nd Int. Conf. on Simulation of Semiconductor Devices and Processes, Swansea: Pineridge, pp. 528-543, 1986

[61] Grubin, H.L., Ferry, D.K. and Gleason, K.R., "Spontaneous oscillations in gallium arsenide field effect transistors", Solid State Electron, 23, pp.157-172, 1980

[62] Wada, T. and Frey, J., "Physical basis of short-channel MESFET operation", IEEE Trans. Electron Devices, ED-26, pp.476-490, 1979

[63] Faricelli, J.V., Frey, J. and Krusius, J.P., "", IEEE Trans. Electron Devices, ED-29, pp.377-388, 1982

[64] Snowden, C.M., "Computer-aided design of MMIC's based on physical device models", IEE Proc., Vol.133, Pt.H, No.5, pp.419-427, 1986

[65] Adler, M.S., "A method for achieving and choosing variable density grids in finite difference formulations and the importance of degeneracy and band gap narrowing in device modelling", Proc. NASCODE 1 Conf., Dublin: Boole Press, pp.3-30, 1979

[66] Adler, M.S., "A method for terminating mesh lines in finite difference formulations of the semiconductor device equations", Solid State Electron, 23, pp.845-853, 1980

[67] Franz, G.A. and Franz, A.F., "Transient 2-D simulation of power thyristors", Proc. Int. Conf. on Simulation of Semiconductor Devices and Processes, Swansea: Pineridge, pp.204-218, 1984

[68] Greenfield, J.A., Price, C.H. and Dutton, R.W., in Process and Device Simulation for MOS-VLSI Circuits, ed. P.Antognetti et al, Amsterdam: Nijhoff, pp.432-480, 1983

CHAPTER 5

NUMERICAL SOLUTION OF THE SEMICONDUCTOR EQUATION FINITE-ELEMENT METHODS

The finite-element method was originally developed for use in structural engineering and may be traced back to work carried out in the early 1940's. Nowadays it is used frequently for analysing structures in civil and mechanical engineering. The finite-element method provides a flexible means of solving the differential equations which constitute the semiconductor equations over complex device geometries. The application of finite-element techniques to semiconductor applications was first reported in the open-literature in the early 1970's [1,2,3]. This method is fundamentally quite different to the finite-difference technique, although it is still necessary to sub-divide the simulation domain into smaller regions - the finite elements (triangles or rectangles). The basis of the finite-element method is to approximate the solution of the differential equation by assuming that the solution follows a simple function over each element. The global solution is then obtained by combining the solutions for the many elements.

An important advantage of the finite-element method over finite-difference schemes is that there is no additional complication introduced by using elements of different sizes. Finite-element methods provide a flexible means of investigating semiconductor devices with non-planar geometries and regions of highly non-linear field and carrier distributions. Continuous functions representing the required solution are assumed to have simple analystic forms within each element (piecewise-linear, cubic). The continuous semiconductor equations are then used to define an equivalent integral formulation. As with the finite-difference method, this method transforms the continuous functions of the semiconductor equations into discretized forms. The principle disadvantages of the finite-element method compared with the finite-difference approach are that initially more effort is required to implement the technique and that the stability and convergence criteria are not as well understood. The finite-element technique applied to semiconductor devices is usually implemented using the Galerkin method of weighted residuals.

5.1. The Finite-Element Method and its Application to Semiconductor Device Simulation

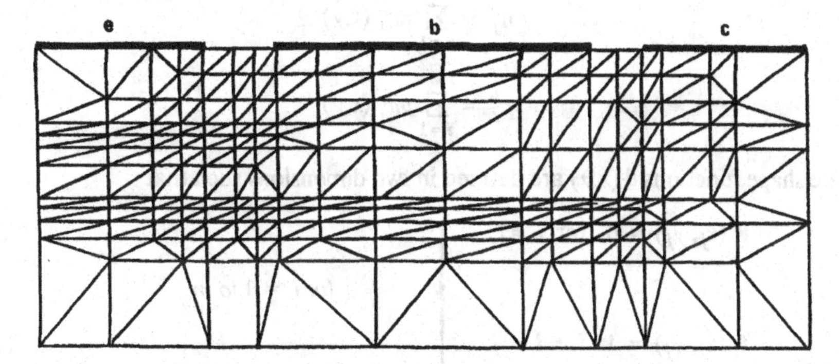

Figure 5.1 Sub-division of the analysis domain for the finite-element method

The domain requiring analysis is sub-divided using piecewise approximations to produce a mesh with m nodes at intersections of the mesh, Figure 5.1. The regions (elements) defined by the mesh are usually triangular or rectangular. The finite-element method produces approximations ψ^h, n^h, p^h to the exact solution for the potential ψ, carriers densities n and p in each of the elements. The approximate solution in each element is known as a partial solution and is determined so that outside the element the contribution to the total approximate solution is zero. The total approximate solution is given by the sum of the partial solutions over all the elements. Hence for n elements,

$$\psi^h = \sum_{j=1}^{n} \psi_j^h \tag{5.1}$$

$$n^h = \sum_{j=1}^{n} n_j^h \tag{5.2}$$

$$p^h = \sum_{j=1}^{n} p_j^h \tag{5.3}$$

It is necessary to obtain a suitable representation for the approximate solution in each element for ψ_j^h, n_j^h and p_j^h. A well behaved (smooth) polynomial is usually chosen. The approximations may then be conveniently formulated by defining 'shape functions' θ_i so that the approximations to ψ, n and p become for m degrees of freedom in each element,

$$\psi_j^h = \sum_{i=1}^{m} \psi_i \, \theta_i \, (x,y) \tag{5.4}$$

$$n_j^h = \sum_{i=1}^{m} n_i \theta_i \, (x,y) \tag{5.5}$$

$$p_j^h = \sum_{i=1}^{m} p_i \theta_i \, (x,y) \tag{5.6}$$

The shape functions $\theta_i(x,y)$ are defined in two dimensions such that

$$\left.\begin{array}{ll} \theta_i \, (x_j, y_j) = 0 \;\; \text{for } i \neq j & (5.7) \\ & \\ \theta_i \, (x_j, y_j) = 1 \;\; \text{for } i = j & (5.8) \end{array}\right\} \quad \textit{for } i = 1 \textit{ to } m$$

Furthermore, the shape functions are non-zero only in the elements adjacent to node j and if j is not on the boundary. On the boundaries $\theta_i(x,y) = 0$. The shape functions are usually linear functions on triangular elements.

The choice of shape function $\theta_i(x,y)$ and the method of dividing up the analysis domain Ω strongly influence the accuracy of solution and the rate of convergence. The order of the polynomial chosen as the shape function determines the number of nodes in each element. For example a quadratic shape function for a two-dimensional solution has six coefficients (x and y) and requires six nodes ($m = 6$). These six nodes could be superimposed on a triangular element. The choice of element shape, polynomial and number of nodes is closely linked. Hence a triangular element could have a linear shape function with three nodes or a quadratic shape function with six nodes. A quadrilateral element with four nodes would be associated with linear shape functions, whilst one with nine nodes would have cubic shape functions. Lagrange polynomials are often used for the construction of shape functions of elements in which only the function values and not derivatives are specified at the nodes. The most commonly used shape function is linear and takes the form [4,5],

$$\theta_i(x,y) = \left[a_i + b_i x + c_i y\right]/2\Delta \tag{5.9}$$

where Δ is the area of the triangle and a_i, b_i c_i are constants determined from the constraints on θ_i given in equations (5.7) and (5.8) which vary from element to element. The gradient of θ_i is constant within an element which can lead to discontinuities in the functions at element boundaries. The Hermite bicubic approximation described by Barnes and Lomax [4] overcomes this difficulty, although it does not allow for actual discontinuities in the device model (such as contacts). A bilinear shape function has been used in the exponentially fitted finite-element method described by Macheck and Selberherr [6].

Substituting equations (5.4) to (5.6) into (5.1) to (5.3) yields the total approximate solutions in terms of the shape functions,

$$\psi^h = \sum_{j=1}^{n}\left[\sum_{i=1}^{m} \psi_i\,\theta_i\,(x,y)\right] \tag{5.10}$$

$$n^h = \sum_{j=1}^{n}\left[\sum_{i=1}^{m} n_i\theta_i\,(x,y)\right] \tag{5.11}$$

$$p^h = \sum_{j=1}^{n}\left[\sum_{i=1}^{m} p_i\theta_i\,(x,y)\right] \tag{5.12}$$

5.1.1. The Galerkin Method

The method of weighted residuals, is usually used to formulate the finite-element equations [7,8]. A popular variation of this technique is the Galerkin method where the shape functions are used to directly determine the weight functions. Formally, the Galerkin method applied in two-dimensions, may be defined for each element in terms of the residual R_i, (error in the solution) in the ith element as

$$\sum_{i=1}^{n}\int R_i\theta_i^j\, dA = 0 \tag{5.13}$$

where n is the number of elements, θ_i^j is the jth shape function in the ith element.

In each element j is chosen so that θ_i^j is non-zero at the node in question. Mathematically, the global set of equations should strictly speaking be regarded as equations at nodes rather than equations in elements (in a similar manner to finite-difference equations).

In order to describe the application of the Galerkin method to the solution of the partial differential equations which constitute the semiconductor equations, consider the generalised set of non-linear partial differential equations,

$$\nabla.\mathbf{F}_i(\mathbf{u},\nabla\mathbf{u}) - c_i(\mathbf{u},\dot{\mathbf{u}}) = 0 \quad i = 1,2,3...n \tag{5.14}$$

Applying the two-dimensional approximation

$$\mathbf{u}(x,y) = \sum_{i=1}^{m} \alpha_i \theta_i(x,y) \tag{5.15}$$

where

$$\alpha_i = \mathbf{u}(x_i,y_i) \tag{5.16}$$

The Galerkin condition requires that

$$R_i(\alpha) = \int_\Omega \theta_i(\nabla.\mathbf{F} - \mathbf{c})d\Omega = 0$$

$$= \int_\Omega \theta_i \mathbf{F}.d\Omega - \int_\Omega (\nabla\theta_i.\mathbf{F} + \theta_i\mathbf{c})d\Omega \qquad i = 1,2,3...m \tag{5.17}$$

Hence the sum over the n finite elements is given by

$$R = \sum_{i=1}^{n}\left[\int_\Omega \theta_i \mathbf{F}.d\Omega - \int_\Omega (\nabla\theta_i. + \theta_i\mathbf{c})d\Omega\right] \tag{5.18}$$

which represents $n \times m$ non-linear equations which is conveniently expressed as

$$\mathbf{R}(\alpha) = 0 \tag{5.19}$$

The Galerkin equations are often solved using the Newton-Raphson iterative technique where,

$$-\frac{\partial \mathbf{r}(\alpha)}{\partial \alpha}\Delta\alpha = -\mathbf{r}(\alpha) = \mathbf{r}(\alpha) \tag{5.20}$$

where

$$\alpha = \alpha + \Delta\alpha \tag{5.21}$$

5.1.2. Finite-Element Semiconductor Equations

The Poisson equation may be re-expressed as,

$$\int_{\Omega} \theta_i (\nabla^2 \psi^h) dV + \int_{\Omega} \theta_i \frac{q}{\varepsilon} \left(N_D - n + p - N_A \right) dV = 0 \qquad (5.22)$$

where V is the volume (for three-dimensions) of the domain.

The finite-element equations are obtained from basic semiconductor equations by multiplying them by θ_i and integrating over the region Ω. This operation corresponds to setting the residuals orthogonal to the set of functions rather than making the residual zero as in finite-difference schemes. Rearranging the equation (5.22) and applying the identity,

$$\nabla.(\theta_i \nabla \psi^h) = \nabla \theta_i . \nabla \psi^h + \theta_i \nabla.(\nabla \psi) \qquad (5.23)$$

and the divergence theorem which states for a generalised function f that the volume integral transforms into the surface integral given by the relationship,

$$\int_{\Omega} \nabla . f \, dV = \int_{S} f . \mathbf{n} \, dA \qquad (5.24)$$

In the case of two-dimensions this has the same effect as applying Green's Theorem. Under these circumstances Poisson equation becomes

$$\int_{\Gamma} \theta_i \, \varepsilon \nabla \psi . \hat{\mathbf{n}} \, dl = \int_{\Omega} \varepsilon \nabla \psi . \nabla \theta_i - q \theta_i \left(N_D - n + p - N_A \right) ds \qquad (5.25)$$

where $\hat{\mathbf{n}}$ is the unit vector normal to the boundary Γ of the domain Ω. In a similar manner, the current continuity equation for electrons after applying the divergence theorem becomes

$$\int_{\Gamma} \theta_i \, \mathbf{J}_n . \hat{\mathbf{n}} \, dl = \int_{\Omega} q \theta_i \frac{\partial n}{\partial t} + \nabla \theta_i . \mathbf{J}_n - q \theta_i G \, ds \qquad (5.26)$$

The finite-element equations are obtained by inserting the approximations ψ^h, n^h and p^h into these equations. The integrals over Γ vanish because of the boundary conditions and definition of the shape functions (for example $\nabla \psi . \hat{\mathbf{n}} = \partial \psi / \partial \hat{\mathbf{n}}$ is the derivative normal to the boundary which is set to zero for free surfaces).

Although Newton-Raphson techniques are often used to solve the non-linear finite-element equations, iterative techniques based on linearising the continuity equations have been very successful in solving these equations [4][9]. In this latter method the potential is calculated from the Poisson solution using carrier densities known at the beginning of the time step. The electric field, carrier velocities and diffusion coefficients are then calculated (assuming that they are constant across the time step). The generation term is evaluated using values of n and p at the beginning of the time step. This process is repeated until the updated values of n and p to calculate the new potentials and other parameters until a converged solution is obtained. Using this iterative technique the Poisson equation is written as,

$$\sum_{j=i}^{m} \left[K_{ij}^{\psi} \psi_j^K - qM_{ij}\left(N_{Dj}^h - n_j^h + p_j^h - N_{Aj}^h\right)\right] = 0 \qquad 1 \le i \le r \tag{5.27}$$

where

$$K_{ij}^{\psi} = \int_{\Omega} \varepsilon \nabla\theta_i . \nabla\theta_j \, ds \tag{5.28}$$

and

$$M_{ij} = \int_{\Omega} \theta_i \theta_j \, ds \tag{5.29}$$

Here r is the number of nodes not lying on contacts ($r < m$). In matrix notation this relationship is given by

$$\left[K^{\psi}\right] \psi^h - q\left[M\right]\left(\mathbf{N}_D^h - \mathbf{n}^h + \mathbf{p}^h - \mathbf{N}_A^h\right) = 0 \tag{5.30}$$

The continuity equation for electrons is expressed in the form

$$q\left[M\right] \frac{\partial \mathbf{n}^h}{\partial t} + \left[K^n\right] \mathbf{n}^h - \mathbf{F}^h = 0 \tag{5.31}$$

where

$$K_{ij}^n = \int_{\Omega} q\nabla\theta_i . \left(\theta_j \mathbf{v}_n^h + D_n^h \nabla\theta_j\right) ds \tag{5.32}$$

where

$$F_i = \int_{\Omega} q\theta_i \, ds \tag{5.33}$$

In the case of holes the continuity equation is,

$$q\left[M\right] \frac{\partial \mathbf{p}^h}{\partial t} - \left[K^p\right] \mathbf{p}^h - \mathbf{F}^h = 0 \tag{5.34}$$

where

$$K_{ij}^{p} = \int_{\Omega} q\nabla\theta_i . \left(\theta_j \mathbf{v}_p^h - D_p^h \nabla\theta_j\right) ds \tag{5.35}$$

Barnes and Lomax have shown that the total current is conserved on an element-by-element basis and that the result is independent of discontinuities in the shape functions [4]. The time derivatives $\partial n^h / \partial t$, $\partial p^h / \partial t$ are usually obtained using finite-difference schemes because of the unwanted coupling between time steps frequently produced by finite-element methods. In the case of equations which are linearised with across one time step, as described above, $[K^n]$ and $\mathbf{F}^n$ are evaluated at the beginning of the time step using values of n and p from the previous time step.

The process of implementing the finite-element technique requires the derivation of separate finite-element equations for each element which are then assembled into a global set of equations. The finite-element matrix equations for the Poisson and current continuity equations are usually solved using direct methods (which normally require large amounts of storage) or an iterative solution such as successive-over-relaxation (SOR). In simulations where a fully coupled simultaneous scheme is not used a block non-linear iteration scheme can be most effective. Block nonlinear iterative algorithms operate on linearised current continuity and a non-linear Poisson equation [9], although convergence is not guaranteed [10,11]. If Newton-Raphson techniques are used to extract $[K^n]$ and $[\mathbf{F}^n]$ at the new time step, the solution time may be lengthier due to the increased effort required to obtain the Jacobian [12]. An advantage of the fully implicit system is that a rapid convergence is usually obtained under all conditions compared with the partially implicit schemes, which may be significant in early stages of the solution (for example the first time step).

More recent finite-element techniques have include the use of different kinds of elements (triangles and rectangles) in the same grid which allow arbitrary subdivision and expansion of grid spacings [6]. This arrangement ensures that only the minimum number of nodes are required. Variable and potential-dependent shape functions have been introduced into some simulations [6]. Special shape functions used in conjunction with Petrov-Galerkin weightings [13] rather than the usual Galerkin weightings promise further improvements in finite-element simulations. These techniques may increase the efficiency of solution and help resolve difficulties in obtaining accurate and stable solutions in narrow interface regions such as in pn junctions.

Mesh-optimization is used in simulations where the carrier and potential functions exhibit sharp variations in isolated regions of the domain while being almost constant in other regions [14]. It also minimises singularities in the solution due to boundary conditions becoming non-analytical in regions where Dirichlet boundary conditions for conducting surfaces abruptly change into Neumann boundary conditions along current-free surfaces [15]. Optimized meshes are also used in regions of the domain where internal boundaries and discontinuities occur. Adaptive meshes are particularly useful in maintaining an accurate solution with the minimum number of elements. The mesh is continually adapted at each time step (or complete iteration cycle for steady-state solutions) as the solution progresses.

Although finite-element schemes usually require fewer nodes than finite-difference schemes, a matrix re-ordering algorithm is often needed to deal with the complex matrix structure which occurs. The re-ordering algorithm minimises computer storage requirements which would otherwise slow down numerical processing [16]. The principal disadvantages encountered in using the finite-element method compared with finite-difference schemes are increased complexity of the programming and the greater density of the matrix equations. This latter factor is particularly significant if higher order approximations and non-uniform meshes are used. However, a non-uniform mesh usually reduces the number of matrix equations required which in turn tends to compensate for the additional complexity of the matrix. Despite these apparent difficulties, the finite-element method offers several important advantages. It is easy to analyse irregular geometries and the computional mesh can be graded to provide a fine mesh in regions of rapid change of variable. This localised mesh refinement is generally easier to implement than in finite-difference schemes. Another advantage of the finite-element method is that it allows high order approximations to be readily produced, which can lead to improvements in accuracy of the solution.

5.2. Examples of Simulations based on the Finite-Element Method

The finite-element method has been applied to a variety of semiconductor devices. Finite-element methods are ideally suited to two- and three-dimensional modelling of non-planar devices because of the adaptable efficient nature of the mesh. However, this method of was not generally applied to semiconductor device modelling until the early 1970's when Barnes and Lomax [4] reported one of the first two-dimensional finite-element simulations which was used to model short gate-length GaAs MESFETs. Since that time, this approach has become comparable in popularity to the finite-difference method and has been extensively used to simulate MOSFET's for VLSI and power applications.

The popularity of finite-element simulation techniques has grown substancially over the last ten years and has been applied to the modelling of a wide variety of semiconductor devices including MOSFETs [17,18,19,20,21], VMOS power FETS [9,22], LDMOS FETs [23,24], BJTs [5,12,25,26], MESFETs [4,27], and non-planar Schottky diodes [28]. Finite-element computional methods have also been applied to simple one-dimensional structures such as p-n junctions [29,30]. Several finite-element modelling software packages have been developed for use in conjunction with device design and optimization. The powerful FIELDAY simulation package, developed by IBM, is based mainly on finite-element techniques and can be used to model a variety of semiconductor devices with particular emphasis on VLSI structures [9,21]. This simulation is currently being used in the computer aided design (CAD) of VLSI devices to predict device behaviour and optimize device designs prior to fabrication.

Bipolar transistors have been modelled using finite-element programs with adaptive meshes. Adachi et al have modelled an npn planar transistor for integrated injection logic (I^2L) applications, Figure 5.2 [5]. Results obtained from this model are shown in Figure 5.3 and include carrier distributions for high injection levels. Adachi et al found that as the injection level increases the rate of convergence of the numerical solutions decreases. This model allows the junction capacitances C_{BE} and C_{BC} to be calculated from the charge Q in the device. as well as the current gain β, cut-off frequency f_T and the static current-voltage characateristics.

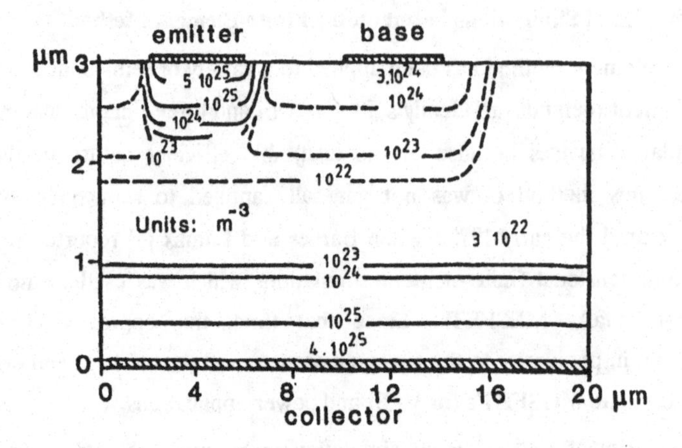

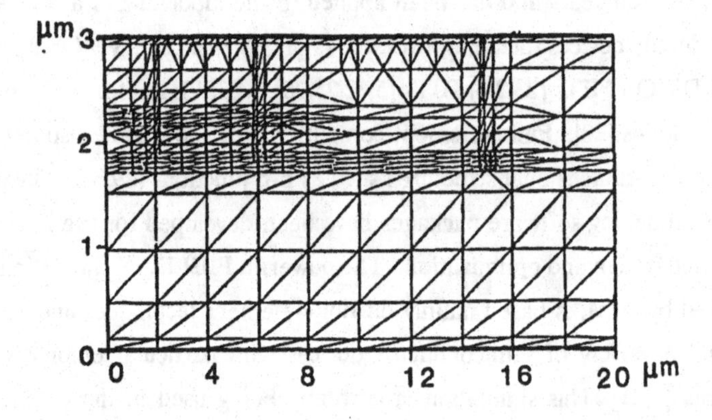

Figure 5.2 Silicon bipolar transistor model showing the finite-element mesh [5]

MOSFET simulations based on finite-element methods have become a popular method of characterising these devices. A simplified two-dimensional diagram of a MOSFET device structure is shown in Figure 5.4. In practice MOSFET's, although surface orientated, tend to be relatively non-planar with respect to the scale of the device.

The structure of a three-dimensional model for small geometry MOSFETs is shown in Figure 5.5. The application of finite-element techniques to non-planar

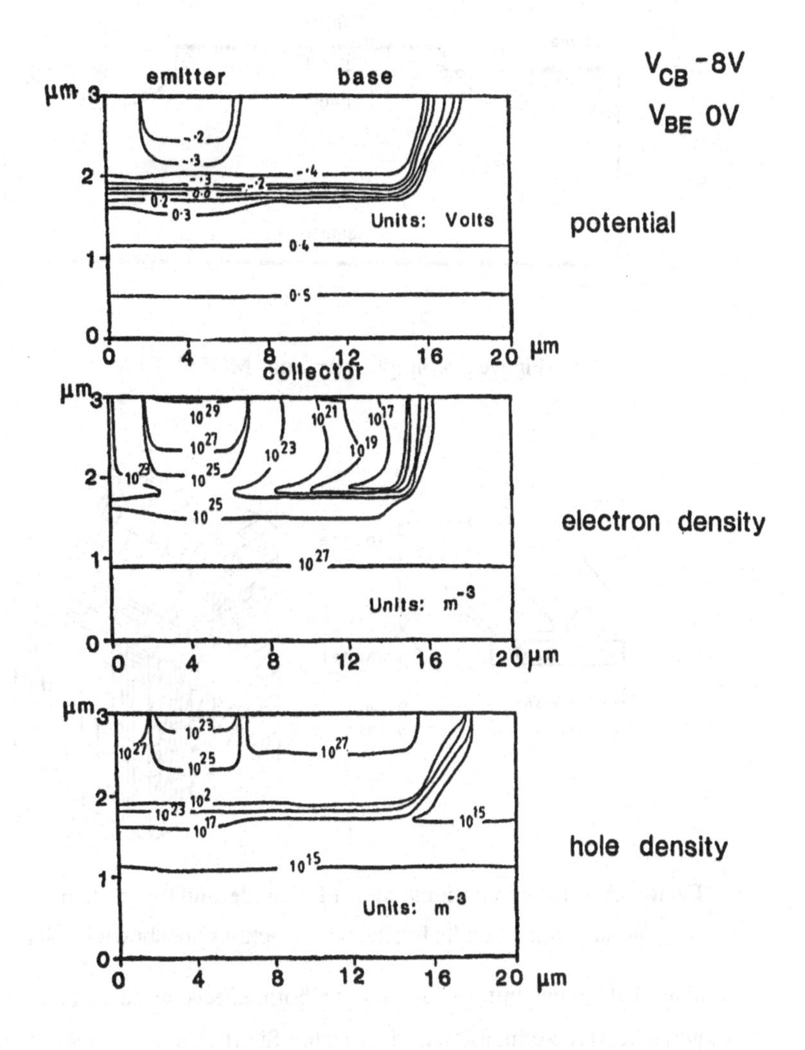

Figure 5.3 Simulation results for a silicon bipolar transistor [5]

devices is well illustrated by Salsburg et al in their modelling of short and narrow silicon MOSFETs [9].

Short- and narrow-channel effects can be modelled independently using this two-

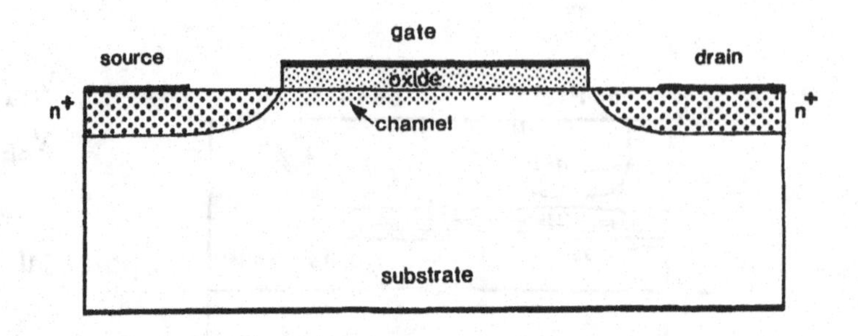

Figure 5.4 Simplified planar MOSFET model

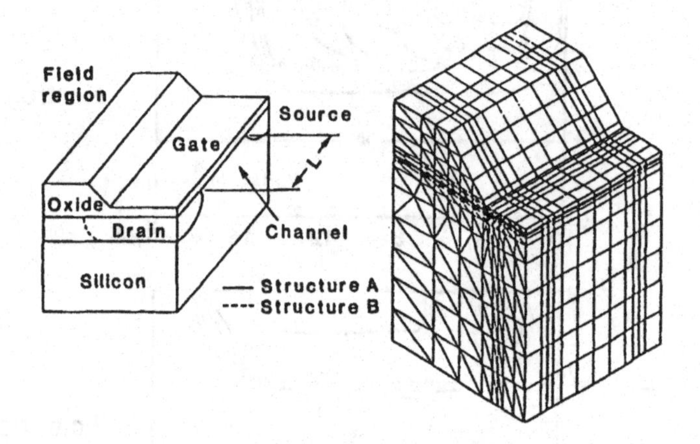

Figure 5.5 Three-dimensional MOSFET model and finite-element mesh [21].
The simulation results for the two structures are shown in Figure 5.6.

dimensional model but to incorporate both effects in the same device model requires a three-dimensional approach. Short-channels cause a decrease in threshold voltage because of the potential barrier lowering due to the proximity of the large drain potential, whilst narrow channels have the opposite effect. Buturla et al, who have simulated devices with gate lengths as short as 0.7 μm and as narrow as 1.5 μm, claim that variations in channel length may contribute more than 50% to the total threshold tolerance of MOSFETs used in VLSI circuits [21].

Simulated subthreshold characteristics were used to define the threshold for a device 1.5 μm long by 1.5 μm wide. It was found that a three-dimensional model is essential to accurately predict the source to substrate bias dependence of the threshold. The results of modelling the surface potential for two different structures are shown in Figure 5.6. In the first structure the source and drain diffusions extend under the field oxide whereas in the second case they terminate at the edge of the field oxide region. This example clearly shows the importance of three-dimensional modelling of VLSI circuit elements in modern semiconductor technology.

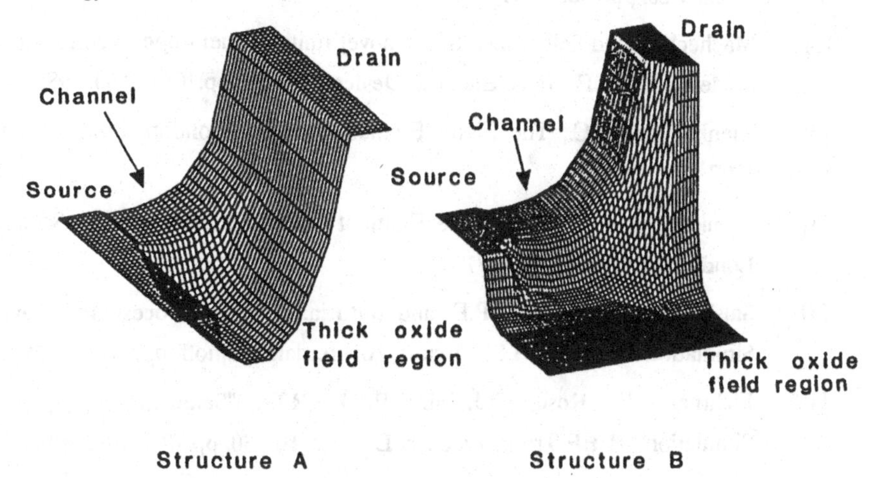

Figure 5.6 Simulation of surface potential for two different MOSFET structures [21]

References

[1] Barnes, J.J, Lomax, R.J. and Haddad, G.I., "Finite element simulation of GaAs MESFET's with lateral doping profiles and submicron gates", IEEE Trans. Electron Devices, ED-23, pp.1042-1048, 1976

[2] Buturla, E.M. and Cottrell, P.E., "Steady state analysis of field effect transistors via the finite element method", Tech. Digest of IEEE Int. Electron Devices Meeting, Washington, pp.51-54, 1975

[3] Wilson, E.A. and Tchon, W.E., "Calculation of transfer potentials using the finite element methods", 1973 SWIEEECO Record of Technical Papers, 25th Annual Southwestern IEEE Conference, Houston, April 1973

[4] Barnes, J.J. and Lomax, R.J., "Finite element methods in semiconductor device modelling", IEEE Trans. Electron Devices ED-24, pp.1082-1089, 1977

[5] Adachi, T., Yoshii, A. and Sudo, T., "Two-dimensional semiconductor analysis using finite-element method", IEEE Trans. Electron Devices, Vol.ED-26, pp.1026-1031, 1979

[6] Macheck, J. and Selberherr, S., "A novel finite element approach to device modelling", IEEE Trans. Electron Devices, ED-30, pp.1083-1180, 1983

[7] Zienkiewicz, O.C., The Finite Element Method, London: McGraw-Hill, 1977

[8] Zienkiewicz, O.C., The Finite Element Method in Engineering Science, London: McGraw-Hill, 1971

[9] Salsburg, K.A., Cottrell, P.E. and Buturla, E.M., in Process and Device Simulation for MOS-VLSI Circuits, Amsterdam: Nijhoff, pp.582-618, 1983

[10] Fichtner, W., Rose, D.J. and Bank, R.E., "Semiconductor Device Simulation", IEEE Trans. Electron Devices, ED-30, pp.1018-1030, 1983

[11] Mock, M.S., "On equations describing steady-state carrier distributions in a semiconductor device", Comm. Pure Appl. Math., Vol.25, pp.781-792, 1972

[12] Hatchel, G.D., Mack, M.H. and O'Brien, R.R., "Semiconductor analysis using finite elements - part II IGFET and BJT case studies", IBM J.Res. Dev. , Vol. 25, No.4, pp.246-260, 1981

[13] Brooks, A.N. and Hughes, T.J.R., "Streamline upwind/Petrov-Galerkin formulations for convection dominated flows with particular emphasis on the incompressible Navier-Stokes equations", Comp. Meth. in Appl. Mech. and Engng., Vol.32, pp.199-259, 1982

[14] Guerrieri, R. and Rudan, M., "Optimization of the finite-element solution of the semiconductor-device Poisson equation", IEEE Trans. Electron Devices, ED-30, pp.1097-1103, 1983

[15] Strang, G. and Fix, G.F., An Analysis of the Finite Element Method, Englewood Cliffs N.J., Prentice-Hall, 1973

[16] Berry, R.D., "An optimal ordering of electronic circuit equations for a sparse matrix solution", IEEE Trans, Circuit Theory, CT-18, pp.40-50, 1971

[17] Barnes, J.J., Shimohigasha, K. and Dutton, R.W., "Short channel MOSFETs in the punchthrough current mode", IEEE Trans. Electron Devices, ED-26, pp.446-453, 1979

[18] Kotecha, H. and Noble, W.P., Proc. IEEE Electron Devices Meeting, pp.724-727, 1980

[19] Chamberlain, S.G. and Husain, A., "Three-dimensional simulation of VLSI MOSFET's", Proc. IEEE Int. Electron Devices Meeting, pp.592-592, 1981

[20] Hatchel, G.D., Mack, M.H., O'Brien, R.R. and Speelpenning, "Semiconductor analysis using finite elements - Part I computational aspects", IBM J. Res. Develop., Vol.25, No.4, pp.232-245, 1981

[21] Buturla, E.M., Cottrell, P.E., Grossman, B.M. and Salsburg, K.A., "Finite-element analysis of semiconductor devices: the FIELDAY program", IBM J. Res. Develop., Vol. 25, No.4, pp.218-231, 1981

[22] Cottrell, P.E. and Buturla, E.M., "Two-dimensional static and transient simulation of mobile carrier transport in a semiconductor", Proc. NASCODE 1, ed B.T.Browne and J.H.Miller, Dublin: Boole Press, pp.31-64, 1979

[23] Darwish, M. and Board, K., "Modelling of LDMOS transistors with ion-implanted and SIPOS surface layers", Proc. Int. Conf. on Simulation of Semiconductor Devices and Processes, Swansea: Pineridge, pp.230-239, 1984

[24] Habib, S.E-D and Embabi, S.H., "Lateral power MOSFET with an improved on-resistance", Proc. 2nd Int. Conf. on Simulation of Semiconductor Devices and Processes, Swansea: Pineridge, pp.594-605,

1986

[25] Hatchel, G.D., Mack, M.H., O'Brien, R.R. and Quinn, H.F., "Two-dimensional finite-element modelling of npn devices", Proc. IEEE Int. Electron Devices Meeting, pp.166-169, 1976

[26] Turgeon, L.J. and Navon, D.H., "Two-dimensional non-isothermal carrier flow in a transistor structure under reactive circuit conditions", IEEE TRans. Electron Devices, ED-25, pp.837-843, 1978

[27] Barnes, J.J. and Lomax, R.J., "Two-dimensional finite-element simulation of semiconductor devices", Electron. Lett., 10, pp.341-343, 1974

[28] Sangiorgi, E., Rafferty, C., Pinto, M. and Dutton, R., "Non-planar Schottky device analysis and applications", Proc. Int. Conf. on Simulation of Semiconductor Devices and Processes, Swansea: Pineridge, pp.164-171, 1984

[29] Please, C.P., "An analysis of semiconductor p-n junctions", IMA Journal of Appl. Mathematics, Vol.28, pp.301-318, 1982

[30] Bettess, P., Greenough, C. and Emson., C., "Speculative approaches to the mathematical modelling of 1-dimensional p-n junctions", Proc. Int. Conf. on Simulation of Semiconductor Devices and Processes, Swansea: Pineridge, pp.247-266, 1984

CHAPTER 6

SEMICLASSICAL TRANSPORT EQUATIONS
HOT ELECTRON EFFECTS

The trend towards smaller semiconductor devices with active regions of the order of one micron or less, has meant that it is necessary to re-examine the applicability of the classical semiconductor equations. The basic classical semiconductor equations described in Chapter 2 are based on the assumption that the carrier energy distribution remains close to its equilibrium form. However, if the electric fields, carrier gradient and current densities in the device become excessively large then non-equilibrium transport conditions will occur. The small geometries encountered in many semiconductor devices frequently have regions of very high electric fields which cause substantial electron 'heating', creating regions where the electrons attain very high energies relative to the equilibrium conditions. In practice these conditions occur in many semiconductor devices used in VLSI, VHSIC and microwave applications.

In 1972, Ruch [1] demonstrated with the aid of Monte-Carlo methods that strongly non-homogeneous electric fields in sub-micron semiconductor devices caused peak velocities far in excess of those attributed to equilibrium velocity-field characteristics. His investigations into short channel GaAs and Si FETs revealed that this velocity overshoot effect dominated the carrier transport process in GaAs and Si FETs with active channel lengths of the order of $1\,\mu m$ and $0.1\,\mu m$ respectively. This non-equilibrium transport process, where the carrier velocity exceeds its steady-state value on a transient basis is also referred to as 'transient carrier transport' (TCT).

Non-equilibrium transport is illustrated in Figure 6.1, which shows the time dependence of the average velocity of electrons injected 'cold' and drifting in a region of uniform electric field for samples of silicon, gallium arsenide and indium

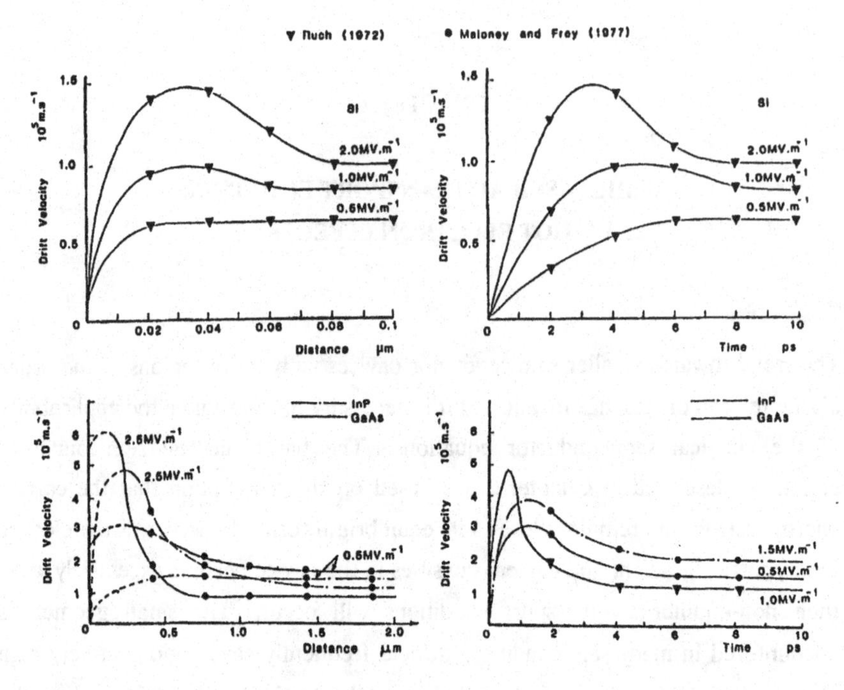

Figure 6.1 Non-equilibrium transport phenomena. Average velocity of electrons injected cold into samples of Si, GaAs and InP with a uniform electric field applied.

phosphide. In all cases the drift velocity overshoots to a value significantly higher than the equilibrium value for the applied electric field. This effect can be attributed to the difference in the momentum and energy relaxation times. The momentum relaxation time is smaller than the energy relaxation time causing the electron distribution to be perturbed first in momentum space. As the energy relaxation becomes effective the distribution function spreads out causing the average drift velocity to decrease. Strong intervalley and acoustic phonon scattering in silicon causes the drift velocity to reach its equilibrium value in less than 0.1 ps. The transient effects in gallium arsenide last substantially longer because of weaker influence of polar optical scattering which is the dominant scattering mechanism. It is evident from this brief discussion that non-equilibrium transport phenomena should be included in models of small devices where the active region is less than 2 μm.

Monte Carlo techniques have been used to provide a valuable insight into the physics of hot electron effects. However, the computation time required to conduct a complete Monte Carlo FET simulation is prohibitively large - requiring in some cases several hours cpu time per bias point [2]. Recent Monte Carlo simulations have benefited from the dramatic improvements in computational power over the last fifteen years, but are still relatively slow. Furthermore, problems in modelling carrier-carrier scattering and quantum effects at the present time limit the usefulness of Monte Carlo models.

Non-equilibrium transport in semiconductors has been successfully modelled using a 'hydrodynamic' semiclassical approach derived from the Boltzmann equation. This method has been mainly applied to the simulation of short gate length MESFET's [3,4,5,6,7,8] although more generalised descriptions are available [6,9,10]. Many of these models have been based on electron temperature in preference to average electron energy [7,9,10].

6.1. The Hydrodynamic Semiclassical Semiconductor Equations

In order to establish a set of transport equations capable of modelling hot electron effects it is useful to re-examine the equations derived from the Boltzmann approximation. Semiclassical transport equations based on either the first three or four moments of the Boltzmann equation are used to derive the carrier concentration, current, average energy and energy flux (if four moments are retained). The coefficients of these equations may be assumed to be a function of average carrier energy. The relationship between these coefficients and the energy are usually determined from steady-state Monte-Carlo calculations and experimental data for homogeneous cases.

The first three Boltzmann moment equations, first introduced in Chapter 2, may be summarised for electrons (neglecting generation and recombination) as,

Particle conservation
$$\frac{\partial n}{\partial t} + \nabla.(n\mathbf{v}) = 0 \qquad (6.1)$$

Momentum conservation
$$\frac{\partial \mathbf{v}}{\partial t} + \mathbf{v}.\nabla\mathbf{v} + \frac{q\mathrm{E}}{m^*} + \frac{1}{m^*n}\nabla\left(nkT_e\right) = -\frac{v(\xi)}{\tau_n(\xi)} \qquad (6.2)$$

Energy conservation
$$\frac{\partial \xi}{\partial t} + q\mathbf{v}E + \mathbf{v}\nabla\xi + \frac{1}{n}\nabla(n\mathbf{v}kT_e) = -\frac{\xi - \xi_o}{\tau_e(\xi)} \qquad (6.3)$$

where k is Boltzmann's constant, ξ is the average electron energy, τ_n and τ_e are the momentum and energy relaxation times, which are themselves functions of the average electron energy. ξ_o is the equilibrium electron energy corresponding to the lattice temperature T_o. The electron energy ξ consists of kinetic and thermal components

$$\xi = \frac{1}{2} m^* v^2 + \frac{3}{2} k T_e \tag{6.4}$$

where m* is the effective mass of the electron and T_e is the electron temperature. If it is assumed that the distribution function is symmetrical in momentum space, as it is for a displaced Maxwellian distribution, then higher order moments of the distribution function vanish. The energy conservation equation (6.3) assumes a displaced Maxwellian distribution, hence heat flow (energy flux) due to the electron gas (which appears as the third moment of the distribution function) is not included in equation (6.3). Blotekjaer points out that this term may be significant for non-Maxwellian distributions and more rigourous derivations include the energy flux term and incorporate the fourth moment equation [10]. For example, an alternative energy transport equation incorporating the energy flux s is [11,12,13],

$$\frac{\partial n\xi}{\partial t} = \mathbf{J}.\mathbf{E} - nB - \nabla.\mathbf{s} \tag{6.5}$$

where

$$\mathbf{s} + \frac{\partial(\tau_h \mathbf{s})}{\partial t} = \mu_\xi n \xi \nabla \psi - \nabla(D_\xi n \xi) \tag{6.6}$$

B is the energy-dissipation factor, μ_ξ, D_ξ and τ_h are the flux mobility, flux diffusivity and high frequency factor respectively. These parameters are taken as functions of average energy ξ and may be defined using integral functions [11]. The terms μ_ξ and D_ξ can be expressed in terms of the electron mobility μ and diffusivity D as follows [12],

$$\mu_\xi = \frac{<\tau_n \varepsilon^2>}{<\tau_n \varepsilon><\varepsilon>} \mu \quad where \quad \mu = \frac{q<\tau_n \varepsilon>}{m<\varepsilon>} \tag{6.7}$$

$$D_\xi = \frac{<\tau_n \varepsilon^2>}{<\tau_n \varepsilon><\varepsilon>} D \quad where \quad D = \frac{kT_e \mu}{q} \tag{6.8}$$

The transport equations outlined above can be usefully simplified by careful consideration of the equation coefficients. In most circumstances the device characteristics are required in the context of time scales associated with circuit time constants in the range 50 ps to 10 ns. This contrasts with time response of the device parameters which lie in the range 0.01 ps to 10 ps. Typically, $\tau_n <\simeq 0.1\,ps$, $\tau_e \simeq 1\,ps$ and $\tau_h = 0.1\,ps$. Hence in many simulations it is common to neglect the products $\tau_n \mathbf{v}.\nabla\mathbf{v}$ and $\partial(\tau_h \mathbf{s})/\partial t$ because they are generally small compared with other terms. For the case of energy transport equations incorporating the energy flux term, such as equation (6.5), the energy transport equation can be written in the form,

$$\frac{\partial n\xi}{\partial t} = \mathbf{J}.\mathbf{E} - nB = \nabla.\alpha[\mu n\xi\mathbf{E} + \nabla(Dn\xi)] \tag{6.9}$$

where

$$\alpha = \frac{<\tau_n\varepsilon^2>}{<\tau_n\varepsilon><\varepsilon>} \tag{6.10}$$

α is usually assumed to be constant, which is a reasonable approximation in the case of power-law scattering. However, power-law scattering implies that the momentum relaxation time is isotropic which is not a good approximation for polar-optical phonon scattering.

For simulations where steady-state solutions are required or transient events with relatively large time scales (>100 ps) are being investigated a quasi-steady state model is adequate. In these circumstances it is possible to neglect the terms $\partial v/\partial t$, $\partial\xi/\partial t$ and $\nabla.n\mathbf{v}$ in the momentum and energy equations (6.2) and (6.3). Furthermore, in cases where carrier-cooling effects are small, such as in MESFETs, it is convenient to neglect the kinetic energy term in equation (6.4) since the thermal energy $3/2\,kT_e$ is typically an order of magnitude greater than the kinetic energy. A quasi-steady-state set of transport equations based on these approximations is,

$$\frac{\partial n}{\partial t} = -\nabla.(n\mathbf{v}) \tag{6.11}$$

$$\mathbf{v} = \frac{\tau_n}{m^*}\left[-q\mathbf{E} - \frac{2}{3}\nabla\xi - \frac{2\xi}{3n}\nabla n\right] \tag{6.12}$$

$$\frac{5}{3}\mathbf{v}.\nabla\xi = q\mathbf{v}E - \frac{\xi - \xi_o}{\tau_e(\xi)} \tag{6.13}$$

Equation (6.12) is equivalent to the more conventional form,

$$\mathbf{v} = -\mu(\xi)\mathbf{E} - \frac{1}{n}\nabla(D(\xi)n) \tag{6.14}$$

taking the charge on the electron as $-q$ Coulombs and where the energy dependent mobility and diffusion coefficients are defined as,

$$\mu(\xi) = \frac{q\tau_n(\xi)}{m^*} \tag{6.15}$$

$$D(\xi) = \frac{kT_e(\xi)\mu(\xi)}{q} \tag{6.16}$$

The definition of electron velocity in the above equations includes the diffusion component and should be distinguished from the drift velocity defined in equation Chapter 2. The mobility and diffusion coefficients for the semiclassical semiconductor equations are functions of average electron energy rather than electric field as in the case of the classical equations derived in Chapter 2. The concept of mobility and diffusion in non-stationary models has attracted some controversy concerning the physical meaning of these terms, although it is accepted that they offer a convenient way of expressing the equations. The electron current density equation which corresponds with the velocity equation (6.12) is

$$\mathbf{J} = -qn\mathbf{v}_n \tag{6.17}$$

The preceeding analysis assumes a single-electron gas model suitable for silicon devices. Compound semiconductors, such as GaAs and InP, with multi-valley band structures would require a more complex treatment. In these semiconductors effective mass variations due to intervalley transfer are important. For the sake of simplicity the analysis is reduced to a two-valley problem, assuming parabolic band structures. A more rigourous treatment would lead to a highly intractable analysis. One possible approximation would be to establish separate transport equations for carriers in the upper and lower valleys, with a suitable treatment of intervalley scattering. Blotekjaer has described a detailed analysis of transport equations for multi-valley semiconductors [10], and has shown that the mobility and diffusion coefficients should be treated as functions of local average velocity of electrons in the lower valley, rather than electric field. Even this two-valley model requires many assumptions - an isotropic semiconductor; nondegeneracy; a displaced Maxwellian distribution; short equivalent valley scattering relaxation times; a constant number of free electrons (no traps); a randomised momentum for intervalley scattering; a scttering rate which depends only on the average kinetic energy of electrons in each valley. In order to simplify the solution of the transport equations it is necessary to assume that the average electron temperature of the electron gases in the upper and lower valleys are the same [7,10].

A more convenient form of the velocity equation can be obtained by re-arranging equation (6.12). A useful interpretation of this expression can be obtained by re-expressing equation (6.12) in terms of electron temperature to yield,

$$v = -\mu(\xi)\mathbf{E} - \left[\frac{k\mu(\xi)}{q}\nabla(T_e) + \frac{kT_e\mu(\xi)}{qn}\nabla n\right] \tag{6.18}$$

A comparison of this equation (in the context of equation 6.17) with the classical current density equation (2.50) reveals that this expansion has an additional term for the spatial variation in electron temperature, which expresses the fact that the most energetic (hot) electrons will diffuse out of the collection. This feature may be significant in devices where there is a high degree of local electron heating such as in Schottky barriers and at the edges of planar contacts [14]. Curtice and Yun have shown that if the temperature gradients are negligible, then the term containing the electron temperature gradient in equation (6.18) may be omitted, yielding an expression identical to equation (6.14) [7]. This approximation is only valid for devices where electron heating is not significant and the gradient term has been shown to be significant in sub-micron structures [8,15]. A consequence of the electron temperature model is that the diffusion coefficient obtained is isotropic rather than anisotropic as predicted by Monte Carlo techniques and measurements. This anisotropic behaviour occurs because the energy is not completely randomized.

A further simplification of the momentum and energy conservation equation (6.2) and (6.3) frequently used in semiconductor device models is to assume that spatial variations are small. This reduces these equations to the following commonly used forms [3,16,17,18],

$$\frac{d(m^*(\xi)\mathbf{v})}{dt} = -q\mathbf{E} - \frac{m^*(\xi)\mathbf{v}}{\tau_n(\xi)} \tag{6.19}$$

$$\frac{d\xi}{dt} = -q\mathbf{E}.\mathbf{v} - \frac{\xi - \xi_o}{\tau_e(\xi)} \tag{6.20}$$

These equations can only strictly be used to model devices where the spatial variation in energy and velocity are small, and hence are not well suited to analysing Schottky barrier or planar devices, where equations (6.2) and (6.3) are better approximations.

In models which assume that the electron energy is a function of electric field alone, which is a reasonable approximation in devices with relatively uniform and slowly varying fields, it is often assumed that spatial and temporal variations are negligible. This leads to a strictly localised model where the momentum and energy conservation equations reduce to the following extremely simple forms,

$$\mathbf{v} = -\frac{q\tau_n(\xi)}{m^*}\mathbf{E} \tag{6.21}$$

$$\xi - \xi_o = -q\tau_e(\xi)\mathbf{E}.\mathbf{v} \tag{6.22}$$

Equation (6.21) is equivalent to the steady-state drift velocity $\mathbf{v} = \mu\mathbf{E}$, which neglects any contribution due diffusion. Substitution of equation (6.21) into (6.22) confirms that the energy ξ is solely a function of electric field, since τ_e and τ_m are functions of the electron energy.

The generation of transport equations for electrons in both the upper and lower valleys leads to a highly complex non-linear model and as previously mentioned it is generally accepted that a single electron gas model is justified [10]. Cook and Frey describe an equivalent single valley model which requires that the fraction of total free carriers in the upper valley and the ratio τ_n/m^* are functions of the localised average energy; the contribution of kinetic energy to the electron temperature is negligible [6]. This latter point implies that velocity overshoot is attributable to the dependence of the momentum relaxation time τ_n on energy. Bozler and Alley equate the thermal energy to the energy separation between the lower and upper valley $\Delta\xi_{UL}$ (0.36 eV for GaAs), the proportion of electrons in the upper valley $G_U(\xi)$ and the total energy ξ [19],

$$\frac{3}{2}kT_e = \xi - G_U(\xi).\Delta\xi_{UL} \tag{6.23}$$

This relationship can be used to modify the momentum conservation equation (6.2) to obtain a model for multi-valley semiconductors, where the velocity is given by

$$v = \frac{\tau_n}{m^*}\left\{q\mathrm{E} + \frac{2}{3}\left[1-\Delta\xi_{UL}\frac{\partial G_U(\xi)}{\partial\xi}\right]\nabla\xi + \frac{2}{3n}\left[\xi - G_u(\xi).\Delta\xi_{UL}\right]\nabla n\right\} \tag{6.24}$$

The highly non-linear behaviour of the energy transport equation means that it is often difficult to solve even the simplified form, equation (6.20). Several approaches have been used to obtain stable, accurate solutions for the energy transport in small geometry devices. Although detailed complete numerical solutions are now available for MESFET's [8,12,15], the problem may be reduced in complexity by identifying regions of hot-electron transport in the device and regions where the carriers do not gain energy from the electric field [6]. In the case of a MESFET, Cook and Frey assume that in regions where carrier heating is significant both the electric field and drift velocity are nearly one-dimensional [6], allowing the energy conservation equation to be approximated (for silicon) as

$$\frac{\partial\xi}{\partial x} = -\frac{3}{5}\left(qE_x + \frac{\xi-\xi_o}{v_x\tau_e\xi_o}\right) \tag{6.25}$$

It is assumed that diffusion effects are negligible compared with the drift velocity contribution in the regions of hot electron transport. Hence, the last term in equation (6.3) can be omitted to yield a one-dimensional transport equation. In the case of silicon the energy transport equation becomes,

$$\frac{\partial\xi}{\partial x} = -\frac{21}{20}qE_x - \frac{9}{20}\left\{\frac{40}{9}\frac{m^*(\xi-\xi_o)}{\tau_n\tau_e} + q^2E_x^2\right\}^{1/2} \tag{6.26}$$

and for GaAs

$$\frac{\partial\xi}{\partial x} = -\frac{21}{20}qE_x - \frac{9}{20}\left\{\frac{8}{3}\left[1+\frac{2}{3}K(\xi)\right]K(\xi)\frac{m^*(\xi-\xi_o)}{\tau_n\tau_e} + q^2E_x^2\right\}^{1/2} \tag{6.27}$$

where

$$K(\xi) = 1-\Delta\xi_{UL}\frac{\partial G_U(\xi)}{\partial\xi} \tag{6.29}$$

These energy transport equations allow the energy ξ to be determined from the electric field subject to appropriate boundary conditions on ξ and E. The determination of ξ and $\tau_n(\xi)$ in these equations is analogous to the calculation of $\mu(E)$ and D(E) in classical models. The momentum and energy relaxation parameters τ_n and τ_e can be obtained from information available from Monte Carlo results for homogeneous samples. In the case of silicon, using a constant effective mass approximation, the steady-state relationships of equations (6.21) and (6.22) for the homogeneous case can be used to obtain τ_n and τ_e,

$$\tau_n(\xi) = -\frac{m^*\mathbf{v}}{q\mathbf{E}} \tag{6.30}$$

$$\tau_e(\xi) = -\frac{\xi-\xi_o}{q\mathbf{E}.\mathbf{v}} = \frac{m^*(\xi-\xi_o)}{q^2E^2\tau_n(\xi)} \tag{6.31}$$

Monte Carlo results and experimental information available on the steady-state velocity-field characteristics allow the relaxation times $\tau_n(\xi)$ and $\tau_e(\xi)$ for silicon to be determined for the homogeneous case. The extraction of these parameters for GaAs is more tedious bacause of the strong dependence of effective mass $m^*(\xi)$ on energy. Steady-state Monte Carlo results describing the electric field dependence of velocity, energy and effective mass for GaAs, obtained from a three-valley Monte-Carlo model, have been used by Carnez et al [3] (Figure 6.2). Electron temperature models which use electric field dependent electron temperature characteristics and temperature dependent energy relaxation times to determine the carrier velocity and relaxation time respectively have been used in other models [15,18,20,21]. Future non-stationary models may have to address the problem of intervalley carrier relaxation which to date has not been included in the models.

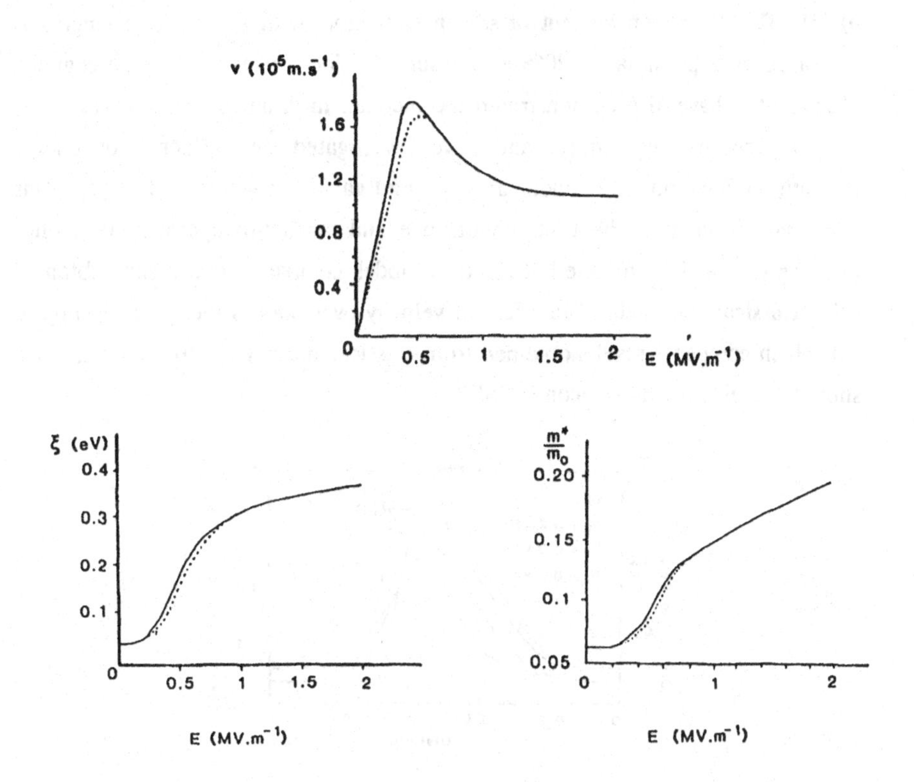

Figure 6.2 Velocity, energy and effective mass characteristics as a function of static electric field

6.2. Examples of Hot Electron Modelling

The development of small-scale devices with active regions of one micron or less has meant that hot electron effects in these devices have been the subject of considerable interest in recent years [3,5,15,16,18,22,23,24,25,26,27]. Hot electron models have tended to concentrate on short gate length MESFET's intended for operation at frequencies upto 70 GHz. Simulations have revealed that velocity overshoot is significant in both Si and GaAs MESFETs where the gate length is less than 0.5 μm. The results show that for the same gate length velocity overshoot is considerably greater in the GaAs FET than in the Si device. The drain saturation current in a 0.25 μm gate length GaAs FET has been shown to be up to three times higher than that predicted by the classical diffusion model (for V_{GS} = 0) [6]. The saturation current of silicon transistors with similar gate lengths is increased by approximately 20% as a result of velocity overshoot in the channel. Carnez et al have also demonstrated the increase in drain to source current I_{DS} due to nonstationary effects and have investigated the influence of energy relaxation effects on transconductance g_m and gain-bandwidth product [3]. This work has shown that the transconductance and maximum operating frequency increase substantially for the hot electron model, compared with results obtained using classical approach. The effect of velocity overshoot is illustrated in Figure 6.3, which compares results obtained from classical and hot electron models for a short channel GaAs and silicon MESFET's.

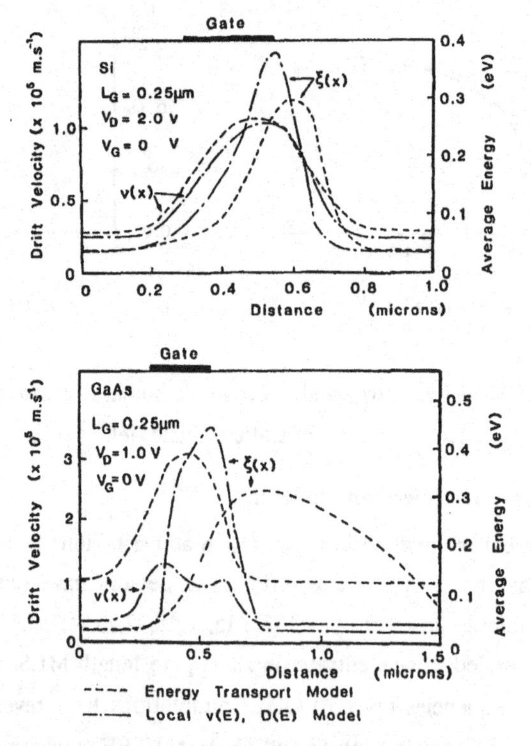

Figure 6.3 Velocity overshoot in GaAs and Si MESFET's

Very short devices may also exhibit ballistic transport. The concept of ballistic transport in very small semiconductor devices has been a topic of considerable debate in recent years. The term 'ballistic transport' refers to the collision-free motion of electrons, which occurs over short time scales. Ballistic transport can occur in devices where the phonon mean free path at the operating temperature is longer than the critical device dimension. The applied voltage must be sufficiently high so that the diffusion term can be neglected and low enough to minimise interband intervalley transitions. Assuming a uniform electric field **E** and ballistic transport conditions, over a period of time τ electrons will move with a velocity given by

$$v = -\frac{q\mathbf{E}\tau}{m^*} \tag{6.32}$$

In practice ballistic transport occurs over such short intervals that it does not lead in itself to large velocities. Over longer periods collision mechanisms limit the velocity to a steady-state drift value. The time scale over which ballistic transport can occur is determined by comparing the average energy acquired by the ballistic electron with the energy at which scattering mechanisms become significant [24].

Ballistic transport in practical devices was originally considered by Shur and Eastman in 1979 in connection with the low temperature (77K) operation of low-power, high-speed devices for logic applications [28]. Near-ballistic transport, which describes transport in the presence of only a few collisions, has been subsequently considered by other researchers [23,29,30]. Near-ballistic transport allows large velocities to be achieved as a result of velocity overshoot. However, ballistic transport (or near-ballistic) cannot by itself lead to overshoot unless the momentum relaxation rate is larger than the energy relaxation rate ($\tau_n \ll \tau_e$) (and is an increasing function of energy), or the semiconductor has satellite valleys with heavier effective mass at low enough energies to have a significant electron population in the steady-state. Hess has shown that for short ($\simeq$0.3 μm) GaAs devices 'near-ballistic' transport occurs only at relatively low voltages ($\simeq$ 0.03 V) and that the transport is 'near diffusion limited' at higher voltages [23]. The transport properties may be enhanced by using materials with a higher optical phonon energy and by screening electron-phonon interaction. Electron-phonon interaction is weaker at high electron densities ($10^{24} m^{-3}$) and close to metal contacts because of remote screening effects. Special device structures with appropriate electric field distributions can be used to enhance velocity overshoot effects. The potential of ballistic transport as a means of enhancing the high frequency performance of transistors has been further investigated by Bozler and Alley who have described a permeable base transistor which may support ballistic transport [19].

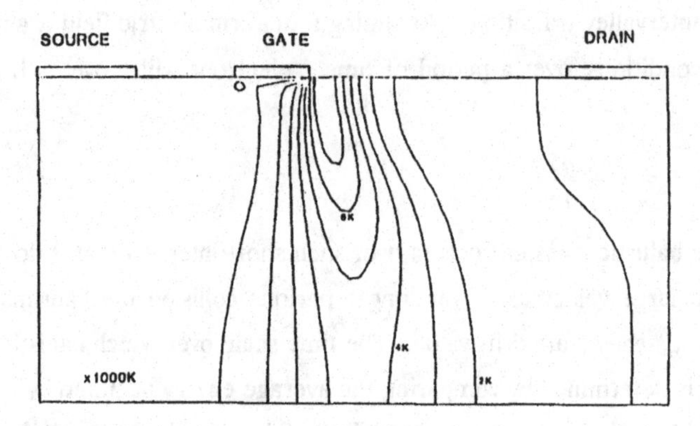

Figure 6.4 Two-dimensional electron temperature distribution in a 0.3 μm gate length GaAs MESFET.

More recent studies of hot electron effects in transistors has shown that the spatial energy dependence can contribute substancially to the electron current density and subsequently modify the simulated characteristics of short gate length MESFET's [8]. These studies have revealed that the average electron temperature distribution in short gate length MESFET's follows a two-dimensional behaviour, Figure 6.4. The current density in the hot electron models is modified by the significant electron temperature gradients and substancial electron temperatures, compared with conventional drift-diffusion models, Figure 6.5.

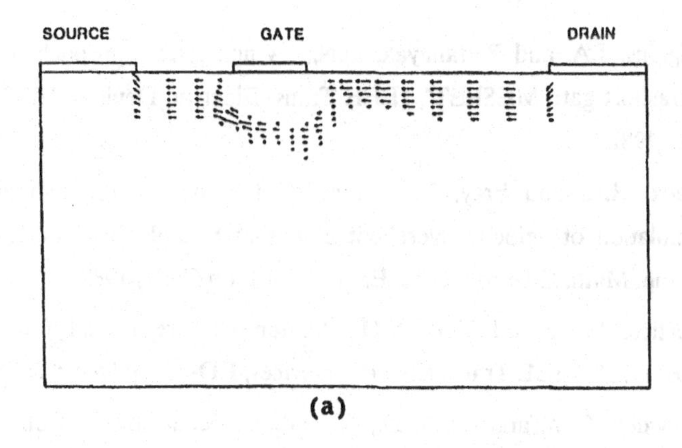

(a)

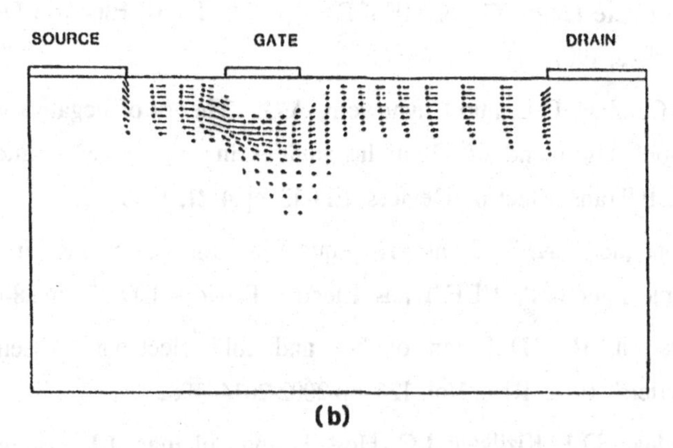

(b)

Figure 6.5 Current density distributions in (a) drift-diffusion MESFET model (b) full electron temperature model of a 0.3 μm gate length GaAs MESFET.

References

[1] Ruch, J.G., "Electron dynamics in short-channel field effect transistors", IEEE Trans. Electron Devices, ED-19, pp.652-659, 1972

[2] Hockney, R.W. and Reiser, M., IBM Res. Rep. R2482, 1972

[3] Carnez, B., Cappy, A., Kaszynski, A., Constant, E. and Salmer, G., "Modeling of a submicrometer gate field-effect transistor including effects of nonstationary electron dynamics", J.Appl. Phys., 51, pp.784-790, 1980

[4] Grubin, H.L., "Switching characteristics of nonlinear field-effect transistors: gallium-arsenide versus silicon", IEEE Trans. Microwave Theory Techniques, MTT-28, pp.442-, 1980

[5] Higgins, J.A. and Pattanayak, D.N., "A numerical approach to modeling ultrashort gate MESFET", IEEE Trans. Electron Devices, ED-29, pp.179-183, 1982

[6] Cook, R.K. and Frey, J., "An efficient technique for two-dimensional simulation of velocity overshoot effects in Si and GaAs devices", Int. J. Comp. Math. Electron. Elec. Engng, Vol.1, pp.65-87, 1982

[7] Curtice, W.R. and Yun, Y.H., "A temperature model for the GaAs MESFET", IEEE Trans. Electron Devices, ED-28, pp.954-962, 1981

[8] Snowden, C.M. and Loret, D., "Two-dimensional hot electron models for short gate length GaAs MESFETs", IEEE Trans. Electron Devices, ED-34, No.2, 1987

[9] McCumber, D.E. and Chynoweth, A.G., "Theory of negative-conductance amplification and of Gunn instabilities in "two-valley" semiconductors", IEEE Trans. Electron Devices, ED-13, pp.4-21, 1966

[10] Blotekjaer, K., "Transport equations for electrons in two-valley semiconductors", IEEE Trans. Electron Devices, ED-17, pp.38-47, 1970

[11] Stratton, R., "Diffusion of hot and cold electrons in semiconductor barriers", Phys. Rev., Vol. 126, pp.2002-2014, 1962

[12] Widiger, D.J., Kizilyalli, I.C., Hess, K. and Coleman, J.J., "Two-dimensional transient simulation of an idealized high electron mobility transistor", IEEE Trans. Electron Devices, Vol. ED-32, pp.1092-1102, 1985

[13] Parker, J.H. and Lowke, J.J, "Theory of electron diffusion parallel to electric fields. 1. Theory", Phys. Rev., Vol. 181, pp.290-301, 1969

[14] Doades, G.P., Howes, M.J. and Morgan, D.V., "Performance of GaAs device simulations at microwave frequencies", Proc. Int. Conf. on Simulation of Semiconductor Devices and Processes, Swansea: Pineridge, pp.353-367, 1984

[15] Snowden, C.M., "Two-dimensional modelling of non-stationary effects in short gate length GaAs MESFETs", Proc. 2nd Int. Conf. on Simulation of Semiconductor Devices and Processes, Swansea: Pineridge, pp.544-559, 1986

[16] Maloney, T.J. and Frey, J., "Transient and steady-state electron transport properties of GaAs and InP", J. Appl. Physics, 48, pp.781-787, 1977

[17] Shur, M., "Ballistic transport in a semiconductor with collisions", IEEE Trans. Electron Devices, ED-28, pp.1120-1130, 1981

[18] Snowden, C.M., "Numerical simulation of microwave GaAs MESFETs", Proc. Int. Conf. on Simulation of Semiconductor Devices and Processes, Swansea: Pineridge, pp.406-425, 1984

[19] Bozler, C.O. and Alley, G.D., "Fabrication and numerical simulation of the permeable base transistor", IEEE Trans. Electron Devices, ED-27, pp.1128-1141, 1980

[20] Wang, Y.C., "A simple analytical model for electron transport in GaAs", Phys. Status Solidi A, Vol. 53, K113, 1979

[21] Loret, D., Baets, R., Snowden, C.M. and Hughes, W.J., "Two-dimensional numerical models for the high electron mobilty transistor", Proc. 2nd Int. Conf. on Simulation of Semiconductor Devices and Processes, Swansea: Pineridge, pp.100-113, 1986

[22] Price, P.J., "Calculation of hot electron phenomena", Solid State Electron., 21, pp. 9-16, 1978

[23] Hess, K., "Ballistic electron transport in semiconductors", IEEE Trans. Electron Devices, ED-28, pp.937-940, 1981

[24] Teitel, S.L. and Wilkins, J.W., "Ballistic transport and velocity overshoot in semiconductors: Part 1 - uniform field effects", IEEE Trans. Electron Devices, ED-30, pp.150-153, 1983

[25] Watanabe, D.S. and Slamet, S., "Numerical simulation of hot-electron phenomena", IEEE Trans. Electron Devices, ED-30, pp.1042-1049, 1983

[26] Cook, R.K., "Numerical simulation of hot-carrier transport in silicon bipolar transistors", IEEE Trans. Electron Devices, ED-30, pp.1103-1110, 1983

[27] Mottet, S. and Viallet, J.E., "Transient hot-electron behaviour in GaAs MESFETs", Proc. 2nd Int. Conf. on Simulation of Semiconductor Devices and Processes, Swansea: Pineridge, pp.68-81, 1986

[28] Shur, M. and Eastman, L.F., "Ballistic transport in semiconductors at low temperatures for low-power high speed logic", IEEE Trans. Electron Devices, ED-26, pp.1677-1683, 1979

[29] Shur, M., "Ballistic and near ballistic transport in GaAs", IEEE Electron Devices Letters, EDL-1, p.147, 1980

[30] Shur, M., "Near ballistic electron transport in GaAs devices at 77°K", Solid State Electron., 24, p11, 1981

CHAPTER 7

SIMULATION OF HETEROJUNCTION DEVICES

There has been considerable interest in recent years in heterojunction devices such as the high electron mobility transistor (HEMT) and heterojunction bipolar transistor (HBT). A heterojunction is a junction between two different semiconductor materials with different energy band-gaps. Heterojunctions are also described as being isotype or anisotype depending on whether the carriers are of the same or different species on each side of the junction. Early research concentrated on heterojunctions of germanium and gallium arsenide [1]. A heterojunction of particular interest at the present time is the GaAs/AlGaAs system used in high frequency HEMT's and HBT's.

Gallium arsenide and aluminium arsenide have very closely matched lattice constants (only 0.12% difference), which allows good quality interfaces to be grown, with very few traps. The composition of AlGaAs can be varied ($Al_xGa_{1-x}As$ - x is the mole fraction), which allows the band-gap to be changed between 1.43eV (GaAs) and 2.15 eV (AlAs). In order for AlGaAs to remain a direct band-gap semiconductor like GaAs, the mole fraction x must be less than 0.45. The detailed physics of heterojunctions is considered elsewhere. However, in order to gain a better understanding of heterojunction operation it is useful to consider energy band diagrams. The exact form of the energy-band diagram for heterojunctions is still the subject of some controversy centred around whether the vacuum level, intrinsic level or conduction band edge respectively should be continuous. Most analyses are based on the Anderson model which assumes a continuous vacuum level. Consider an abrupt p-N heterojunction (the upper case N indicates the wide band-gap material) shown in Figure 7.1. Band-bending occurs because of charge redistribution caused by the requirement to maintain continuity of the Fermi-level. A similar bending occurs in homojunctions. However, as the band-gaps are different for the two materials in the case of the heterojunction, a discontinuity occurs at the interface producing a 'spike' and 'notch' in the conduction band.

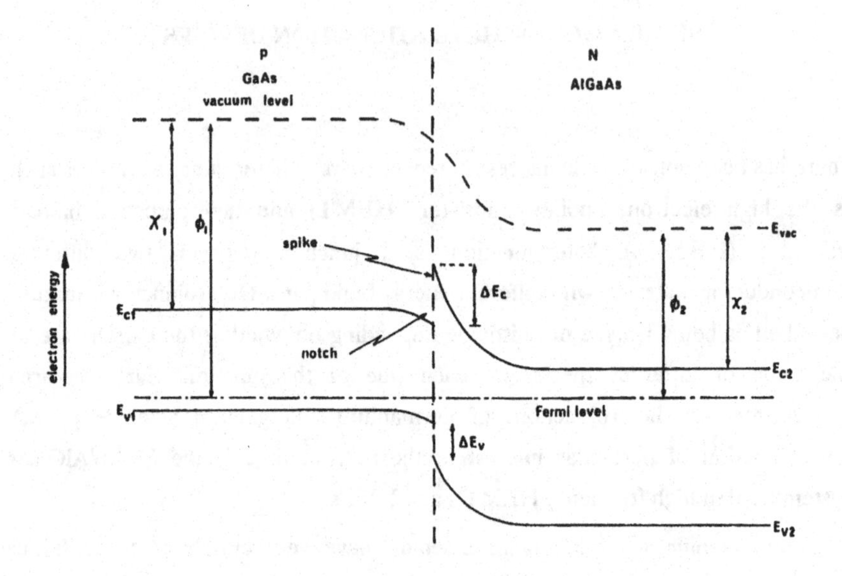

Figure 7.1 Energy band diagram for a p-N heterojunction in equilibrium

In order to simulate heterojunctions it is necessary to take account of several parameters which differ from the homojunction and unipolar devices considered earlier in this text. In particular the band-gap, electron affinity, permittivity, mobility and diffusion coefficients have a positional dependence in heterostructures. Furthermore, there is a built-in field due to changes in the band-gap and significant interface recombination in the vicinity of the notch.

Classical and semiclassical models have been applied to the simulation of heterostructure devices. An analytical model, suitable for modelling HEMT's, has been described by Delagebeaudeuf and Linh [2]. A modified form of the one-dimensional bipolar simulation SEDAN has been used to model GaAs/AlGaAs heterojunction bipolar transistors, predicting cut-off frequencies in excess of 100 GHz [3]. Several heterojunction device models have been based on the formulation of Lundstrom and Schuelke [4]. A further model based on Fermi-Dirac statistics, suitable for heavily doped semiconductors (the base regions in HBT's have doping levels of the order of $10^{24} m^{-3}$) has also been described by Lundstrom and Schuelke [5]. This model has been used to investigate the influence of the conduction band spike in HBT's and the gate bias dependence of the two-dimensional electron gas in HEMT's.

7.1. Semiconductor Equations for Heterojunctions

In a heterojunction device the relative permittivity ε_r varies according to the position in the device and the composition of the material. This implies that Poisson's equation must be modified to take account of the inhomogeneous nature of the material. Re-writing Poisson's equation as

$$\nabla.(\varepsilon_o \varepsilon_r \mathbf{E}) = \rho \tag{7.1}$$

then

$$\mathbf{E}.\nabla \varepsilon_o \varepsilon_r + \varepsilon_o \varepsilon_r \nabla.\mathbf{E} = \rho \tag{7.2}$$

For the sake of simplicity the analysis will be confined to one-dimension, where,

$$E_x \frac{d\varepsilon_o \varepsilon_r}{dx} + \varepsilon_o \varepsilon r \frac{dE_x}{dx} = \rho(x) \tag{7.3}$$

substituting for the electric field in terms of the potential ψ,

$$\frac{d^2\psi}{dx^2} = \frac{q}{\varepsilon_o \varepsilon_r}\left(N_D - n + p - N_A\right) - \frac{1}{\varepsilon_o \varepsilon_r}\frac{d\varepsilon_o \varepsilon_r}{dx}\frac{d\psi}{dx} \tag{7.4}$$

The final term on the right-hand side of equation (7.4) is zero for devices fabricated from homogeneous material where $\nabla \varepsilon_o \varepsilon_r = 0$.

The current density equations must also be re-formulated to take account of the modified energy-band structure in the non-homogeneous device. Under conditions of thermal equilibrium, for a non-degenerate semiconductor, the electron and hole current densities may be expressed in terms of the gradients of the quasi-fermi levels E_{fn} and E_{fp},

$$\mathbf{J}_n = -n\mu_n \nabla E_{fn} \tag{7.5}$$

$$\mathbf{J}_p = -p\mu_p \nabla E_{fp} \tag{7.6}$$

The electron and hole Fermi energy levels are given by

$$E_{fn} = q\psi - (\chi - \chi_r) + kT \exp\left(\frac{n}{N_c}\right) \tag{7.9}$$

$$E_{fp} = -q\psi - (\chi - \chi_r) + E_g - kT \exp\left(\frac{p}{N_v}\right) \tag{7.10}$$

where χ is the electron affinity, χ_r is the reference level, E_g is the band-gap, N_c and N_v are the conduction and valence band density of states. Substituting for E_{fn} and E_{fp} in equations (7.5) and (7.6) yields the modified current density equations,

$$J_n = \mu_n n\left(q\mathbf{E} - \nabla\chi + \frac{kT}{n}\nabla n - \frac{kT}{N_c}\nabla N_c\right) \tag{7.11}$$

$$J_p = \mu_p p\left(q\mathbf{E} - \nabla\chi - \nabla E_g + \frac{kT}{p}\nabla p - \frac{kT}{N_v}\nabla N_v\right) \tag{7.12}$$

In the homogeneous case, equations (7.11) and (7.12) reduce to the familiar current density equations in Chapter 2. The heterostructure equations outlined above require the following implicit assumptions. Firstly, all impurity sites are assumed to be completely ionised. It is assumed that the spatially dependent effective mass remains a valid concept and that the Anderson electron affinity rule may be applied. These two assumptions require the grading to be 'sufficiently gradual' that equations (7.5) and (7.6) remain valid. It is assumed that there is no interfacial charge at the heterojunction due to lattice mismatch and that the doping is sufficiently low to avoid degeneracy. Boltzmann statistics are assumed to apply to the carrier transport process. The influence of Fermi-Dirac statistics on the Einstein relationship has been accounted for in the work of Lundstrom and Schulke by adding an extra term to equations (7.5) and (7.6) [5]. However, if the doping is far below degeneracy this additional term has negligible influence. It is assumed that band-gap narrowing does not occur Finally, it is assumed that the minority carriers and majority carriers of the same species have the same mobility. The characteristics of heterojunctions have been investigated using both closed-form and numerical modelling techniques. For example the influence of the grading profile across the heterojunction on the I-V characteristics of the junction have been investigated using a detailed numerical model [6]. The simulated characteristics for abrupt and linearly graded (over 500Å) junctions are compared with those of a similar homojunction in Figure 7.2.

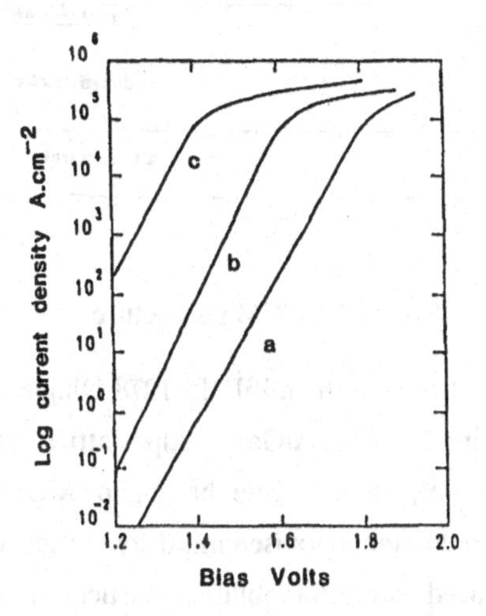

Figure 7.2 Simulated current-voltage characteristics for abrupt and graded heterojunctions

7.2. High Electron Mobility Transistors

The high electron mobility transistor or HEMT, also known as the TEGFET (two-dimensional electron gas FET) or MODFET (modulation doped FET), has undegone rapid development over the past few years. The first discrete devices began to appear in 1980 [7,8]. Since that time they have been incorporated into a variety of digital and analogue integrated circuits, operating at microwave frequencies.

HEMT's are fabricated using a GaAs/AlGaAs heterostructure, Figure 7.3. Electrons are attracted towards the narrow band-gap material and accumulate at the interface between the two materials as a result of the band bending. In 1969, Esaki and Tsu made the important discovery that ionised donor impurities and free electrons could be spatially separated using a superlattice structure to reduce

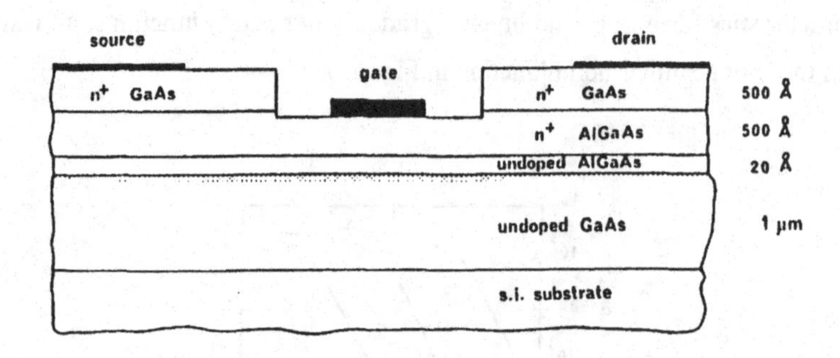

Figure 7.3 HEMT structure

the coulombic interaction between them [9]. In 1978 Dingle and co-workers grew the first modulation-doped GaAs/AlGaAs superlattice in which the donor impurities were present only in the wide band-gap AlGaAs layer [10]. This resulted in a substancial reduction in ionised impurity scattering in the GaAs layers which led to much enhanced electron mobilities, particularly at low temperatures. The electrons from the ionised donors in the AlGaAs transfer into the GaAs and form a conducting layer which is relatively free of ionised impurity scattering sites. The potential well formed by the notch is normally narrow enough to have well defined quantised energy levels in the longitudinal direction and is usually treated as a two-dimensional electron gas. Undoped AlGaAs spacer layers were introduced to further reduce ionised impurity scattering. Mobilities as high as 240 $m^2V^{-1}s^{-1}$ at 2K have been recorded (300 times higher than that of intrinsic bulk GaAs). This very high mobility has led to the highest recorded transconductances being observed (500 mS/mm at 77K) and much improved high frequency performance. HEMT's also exhibit excellent low noise performance under small-signal conditions. HEMT's have been used to fabricate 4k static RAMs which have demonstrated an access time of 2 ns at 77k [11]. They have also been used to produce high performance millimetre wave amplifiers with less than 3.4 dB noise figure and greater than 5.6 dB associated gain at 44 GHz [12].

7.2.1. Closed-Form Models

Analytic closed-form models have been developed to characterise heterostructure devices [2,13,14,15]. These models are usually based on a one-dimensional treatment of the two-dimensional electon gas layer (2-DEG). One particularly useful dc model [16] treats the device in three sections. The model assumes that there is a linear region in which the carrier velocity is proportional to the applied electric field; a second region in which the carrier velocity is constant and where the two-dimensional electron gas is under the influence of the Schottky barrier gate; and a third section beyond the edge of the gate where the carriers are assumed to remain at the saturation velocity. It has been shown that for fields below the critical value, the potential in the channel can be described as a function of distance x from the source edge of the gate towards the drain [16],

$$V(x) = V_G - V_P - \left[\left(V_G - V_P - I_D R_S\right)^2 - \frac{2d_2 I_D x}{\mu_n W_G \varepsilon_o \varepsilon_{r2}}\right]^{1/2} \qquad (7.17)$$

where V_G is the gate voltage, I_D the channel current, V_P the pinch-off voltage, R_S the source resistance, d_2 the gate to 2DEG spacing, μ the electron mobility, W_G the gate width and $\varepsilon_o\varepsilon_{r2}$ the AlGaAs permittivity.

In the second of the two regions, below the gate, it is necessary to solve the two-dimensional Poisson equation.

$$\frac{\partial E_x}{\partial x} + \frac{\partial E_y}{\partial y} = \frac{qN_s}{d_2 \varepsilon_o \varepsilon_{r2}} \qquad (7.18)$$

where N_s is the net sheet charge (positive). This equation may be reduced to a second-order differential equation in E_x which is solved to yield a potential distribution which is a sinh function of x at the drain edge of the gate.

The final section, where the electrons travel at saturated velocity, has a solution which assumes that $\partial E_y/\partial y = 0$ because the E field is assumed to be predominantly longitudinal beyond the drain edge of the gate. The solution is therefore a linear function of E_x.

The drain current is obtained by comparing the total potential across the three regions woth the applied drain-source voltage. The 2DEG sheet carrier concentration N_s is assumed to be a linear function of gate voltage [17]. This involves calculating N_s by solving Poisson's equation in the doped AlGaAs layer, and from an appropriate solution of Schrödinger's equation assuming a triangular well. The fermi-level position is then adjusted in an iterative process until the two values of N_s agree.

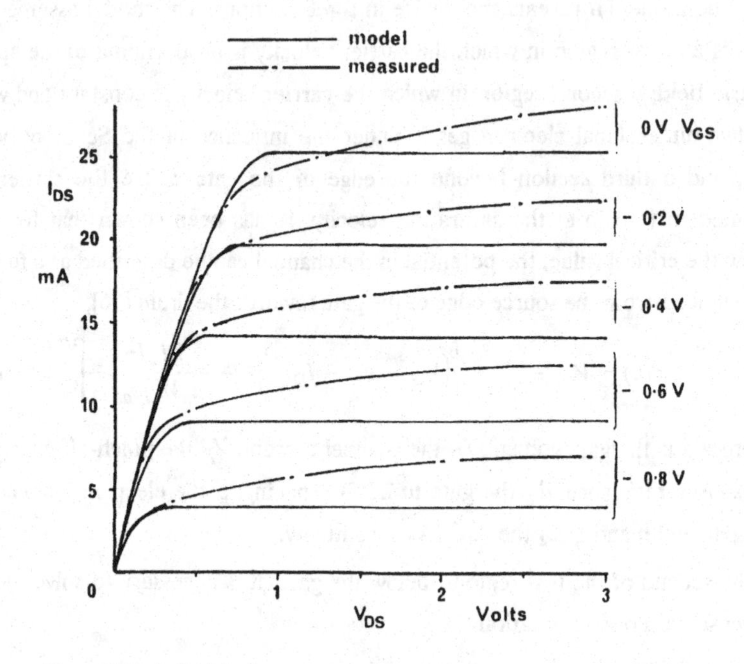

Figure 7.4 Calculated current-voltage charactersitics of a HEMT

The calculated I-V characteristics of a 0.7 μm gate length HEMT are shown in Figure 7.4. The characteristics of a fabricated device are included for comparison. The closed-form analysis provides good agreement between the measured and calculated saturation current, pinch-off and knee voltages. However, the model cannot account for the finite incremental drain conductance in the saturation region found in practical devices. Improved results can be obtained from detailed numerical models, although the relative speed advantage of the above closed-form analysis may be advantageous in many applications.

7.2.2. Numerical Models

The inability of most closed-form models to model accurately the potential, carrier, and field distributions in short gate length HEMT's and account for substrate and AlGaAs conduction processes has led to interest in developing numerical simulations. At the time of writing, there are relatively few numerical HEMT simulations available, although the frequency at which they are being reported is increasing [18,19,20].

The majority of numerical models developed to date are based on the macroscopic semiconductor equations described previously. Researchers have investigated both the simplified drift-diffusion approach and the more complete energy transport models [19,20,21]. Monte Carlo simulations have been used to determine the mobility, longitudinal and transverse diffusion coefficients for both the bulk and quantum well [22]. The velocity-field characteristic of bulk GaAs and AlGaAs obtained from Monte Carlo simulations are shown in Figure 7.5.

Hot electron effects have been shown to be important in HEMT operation [20,21]. Simulations generally consider a simplified device geometry consiting primarily of a AlGaAs/GaAs interface, rather than the complex multi-layer MBE structures currently popular in research devices. Electron transport in the HEMT can take place in both the bulk GaAs (and in some cases the AlGaAs) and in the quantum well. In the region near the source where the fields are small the conduction process may be attributed mainly to the lowest quantum sub-band. In contrast, in the pinchoff region at the drain end of the gate, the electron current will be principally in the bulk. Using these assumptions Widiger's model considers only the lowest sub-band and bulk transport.

The semiclassical transport equations for electrons in the quantum well are taken as similar to those in the bulk, described in the previous chapter with the additional constraint that electrons can transfer into the bulk by means of coupling terms introduced into the transport equations. Hence at the interface,

$$\frac{\partial n_I}{\partial t} = \frac{1}{q}\nabla.\mathbf{J}_I + \frac{1}{q}J_c \tag{7.19}$$

$$\frac{\partial n_I \xi_I}{\partial t} = \mathbf{J}_I\mathbf{E}_I - n_I B_I + \nabla_I.\alpha[\mu_I n_I \xi_I \mathbf{E}_I + \nabla_I(D_I n_I \xi_I)] - S_c \tag{7.20}$$

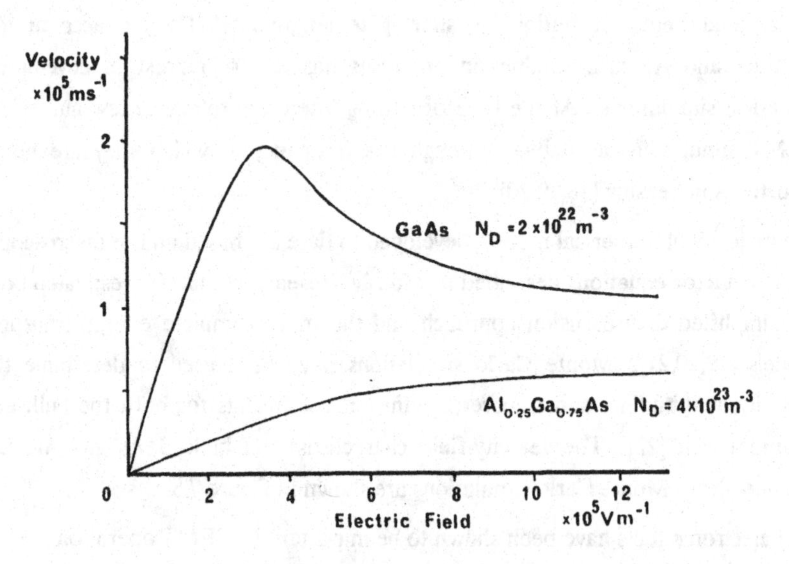

Figure 7.5 Velocity-field characteristic of AlGaAs obtained from Monte Carlo simulations

where the subscript I refers to parameters defined in the interface. The terms J_c and S_c couple the sub-band to other sub-bands or the bulk system. For the case of only two systems which require coupling, the lowest sub-band and the bulk at the interface, the coupling coefficients are determined by the current and energy flux components perpendicular to the interface,

$$J_c = \mathbf{J}_{norm}\Big|_{interface} \tag{7.21}$$

$$S_c = \mathbf{S}_{norm}\Big|_{interface} \tag{7.22}$$

These coupling coefficients guarantee conservation of current and energy across the interface. It is also necessary to define the rate of transfer between the two systems. Widiger et al [20] assume that the relative concentrations of the quantum system and the adjoining edge of the bulk system follow a localised quasi-equilibrium. This is achieved by eliminating the quasi-Fermi energy from the respective electron concentrations n_{int} at the interface and n_I,

$$n_I = N_{I_c} ln\left[1 + \frac{n_{int}}{N_c} \exp\left(\frac{E_1 - E_0}{kT_e}\right)\right] \tag{7.23}$$

where E_0 and E_1 are the minimum energies of the first two sub-bands, T_e is the average electron temperature. N_c and N_{I_c} are the effective density of states of GaAs in three and two dimensions respectively (in the bulk and two-dimensional electron gas respectively). Note that the energy levels are denoted E_x ($x = 0, 1..$) rather than the average energy notation ξ, which should not be confused with the electric field **E**. It is assumed that the potential of the bulk system at the edge of the quantum well is at the second sub-band energy and that the bulk system is not degenerate. The quasi-equilibrium requires that the average energies of the two systems are equal,

$$\xi_I = \xi\Big|_{int} = \frac{3}{2} kT_e \tag{7.24}$$

It is important to appreciate that equation (7.23) indicates that at high electron temperatures or reduced quantum energy spacing, the bulk concentration at the edge increases relative to the quantum well concentration. Although n_I approaches zero in the limit the ratio of the two concentrations is finite. In these circumstances the quantum well width tends towards infinity and the lowest sub-band becomes part of the bulk band structure, leading to a three-dimensional rather than two-dimensional system. Equation (7.23) is modified for cases where the average electron temperature (and hence the electron energy) approaches the energy level separation ($E_1 - E_0$), so that,

$$n_I = N_{I_c}\left[1 - \exp\left(\frac{E_0 - E_1}{kT_e}\right)\right] . ln\left[1 + \frac{n_{int}}{N_c} \exp\left(\frac{E_1 - E_0}{kT_e}\right)\right] \tag{7.25}$$

The calculation of E_0 and E_1 strictly requires a self-consistent solution of Schrödinger's equation. Widiger et al have described efficient approximate solutions for E_0 and E_1 which take account of the main features of the quantum-mechanical solution [20].

An alternative simplified inreface expression has been described by Loret and Snowden [21], based on the thermionic emission model of Schuelke and Lundstrom [23]. In this model the current density normal to to the interface at the abrupt heterojunction is replaced by the expression,

$$J_{norm} = -qS_n\gamma_n\left[n^- - n^+\exp\left(\frac{-\Delta V_n}{kT}\right)\right] \qquad (7.26)$$

where S_n is the interface velocity, n^- and n^+ are the electron densities near the junction on the AlGaAs and GaAs sides respectively. Tunnelling is taken into account by including the parameter γ_n, obtained for a triangular barrier [24]. ΔV_n is the discontinuity in the energy band parameter,

$$\Delta V_n = \Delta\chi + kT\log\left[\frac{N_c^+}{N_c^-}\right] = kT\log\left[\frac{F_1/2(n_c)}{\exp(n_c)}\right] \qquad (7.27)$$

where χ is the electron affinity, N_c the density of states (the + and - superscripts indicating the GaAs and AlGaAs sides of the interface). The discontinuity in the conduction band χ is taken to be 62% from more recent investigations. The third term on the right-hand side of equation (7.27) is a correction factor to allow Fermi-Dirac statistics to be accounted for [23]. This model does not take into account sub-band splitting in the two-dimensional electron gas. This leads to a difference of the order of 20% between results predicted by exact quantum mechanical calculations and the semiclassical approach using Boltzmann statistics. The inclusion of Fermi-Dirac statistics reduces this error to a few percent.

The Poisson equation is solved for the HEMT in the usual manner described previously subject to the the provision of suitable interface boundary conditions. The interface between the GaAs and AlGaAs has a dielectric discontinuity and an interface charge n_I. The potential must be continuous so that

$$\psi\Big|_{int^+} = \psi\Big|_{int^-} \qquad (7.28)$$

where the + and - superscripts indicate the GaAs and AlGaAs sides respectively. The perpendicular displacement difference equals the interface charge,

$$\varepsilon_{ins}\frac{\partial\psi}{\partial r_{norm}}\Big|_{int^-} - qn_s = \varepsilon_{sc}\frac{\partial\psi}{\partial r_{norm}}\Big|_{bulk^+} \qquad (7.29)$$

Also in Widiger et al's model the potential at the second quantum level at the edge of the bulk system is related to that in the AlGaAs by

$$q\psi\Big|_{bulk^+} = q\psi\Big|_{int^-} - E_1 \tag{7.30}$$

The effects of surface charge on the free-surface (top) of the device can be included in the boundary condition. By taking the perpendicular field in the vacuum to be zero (outside the device), a surface boundary condition of the following form is derived [20],

$$-qn_{ss} = \epsilon_{ins} \frac{\partial\psi}{\partial r_{norm}}\Big|_{vac^+} \tag{7.31}$$

where n_{ss} is the surface state concentration which is a function of the surface states. n_{ss} may be estimated by taking it as a linear function of position where the end values are determined at the edges of the gate and the source or drain contact. Alternatively, it may be taken to be a constant value determined from the one-dimensional equilibrium solution. Widiger et al discuss the relative merits of these two approaches and conclude that the effects of surfaces may be significant for HEMT's.

Examples of the electron density and electron energy for a 0.7 μm gate length HEMT obtained from a numerical simulation are shown in Figure 7.6. The solutions were obtained using finite-difference schemes with a variable mesh (strongly refined in the region of the GaAs/AlGaAs interface). The electrical parameters g_m, C_G and f_T obtained from the simulation are shown in Figure 7.7. The transconductance g_{m0} was obtained using the incremental relationship,

$$g_{m0} \simeq \frac{\Delta I_D}{\Delta V_{GS}}\Big|_{V_{DS} = const} \tag{7.32}$$

The gate capacitance C_G was calculated from,

$$C_G \simeq \frac{\Delta Q}{\Delta V_{GS}}\Big|_{V_D = const} \tag{7.33}$$

where the charge is obtained by integrating over the device,

$$Q = \int\int qn \, dx \, dy \tag{7.34}$$

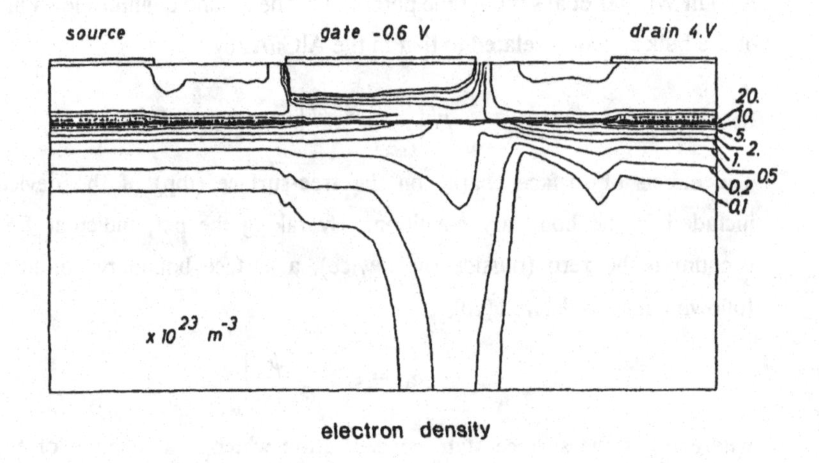

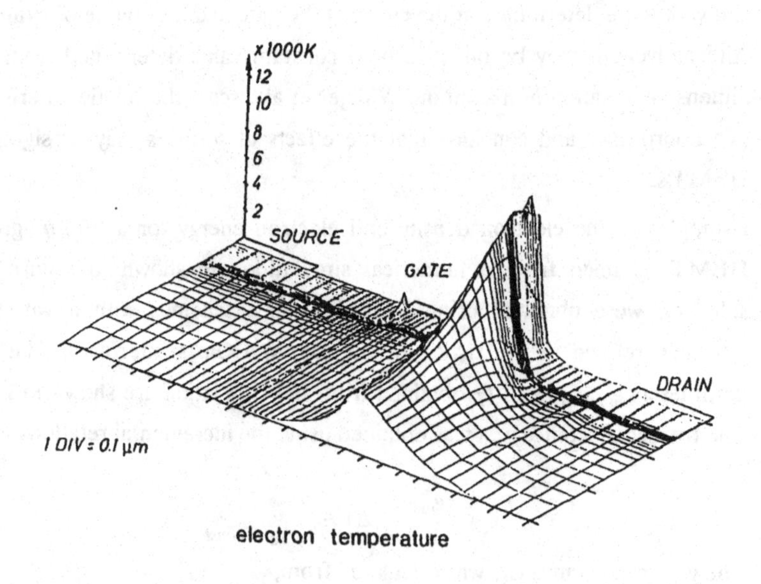

Figure 7.6 Electron density and electron temperature distributions obtained from the simulation of a 0.7 μm gate length HEMT.

The unity gain cut-off frequency f_T is then given by

$$f_T = \frac{g_m}{2\pi C_G} \tag{7.35}$$

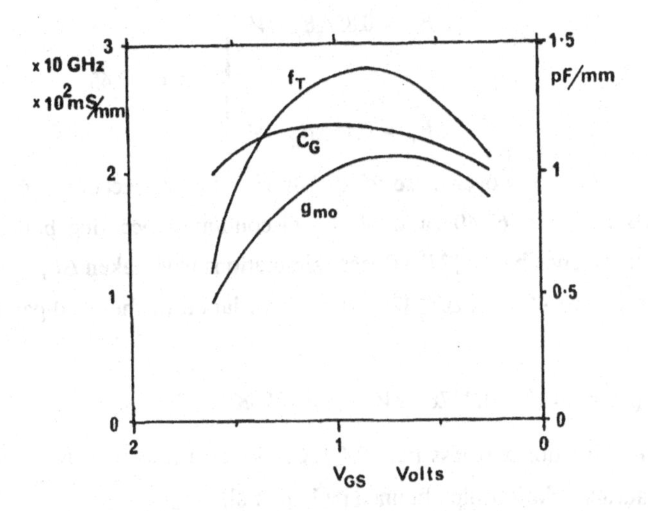

Figure 7.7 Electrical parameters obtained from HEMT simulation

7.3. Heterojunction Bipolar Transistors

The physical modelling of heterojunction bipolar transistors (HBT's) has received less attention than that of HEMT's, even though interest in the use of this type of transistor is growing for high speed logic and microwave-optical applications. The key feature of HBT operation is the confinement of holes to the base region, which allows high injection efficiencies. The emitter-base junction is fabricated as a heterojunction,.with an AlGaAs emitter and GaAs base. The potential barrier at the junction opposes hole injection into the emitter. This means that it is not necessary to have an emitter that is more heavily doped than the base region to achieve high injection ratios, as in homojunction transistors. The base doping can be increased to decrease the base resistance, reducing the junction charging times and so increasing the cut-off frequency f_T and maximum frequency of oscillation f_{max}. It has been suggested that cut-off frequencies in execess of 100 GHz may be achieved for this type of transistor.

Numerical simulations based on the modified semiconductor equations discussed above have been reported [3,6,25]. The energy band structure is frequently based on results of Dingle [26], where

$$\left.\begin{aligned} \Delta E_c &= 0.85 \Delta E_g \ eV \quad (7.36) \\ \Delta E_v &= 0.15 \Delta E_g \ eV \quad (7.37) \end{aligned}\right\} 0 < x < 0.45$$

There is some controversy over these relationships and more recent measurements point towards nearer a 60:40 ratio of the discontinuity occuring between the conduction and valence bands [27]. Recent simulations have taken ΔE_c as 60% of ΔE_g, with ΔE_v being 40% of ΔE_g [21,29]. The variation of the band-gap E_g may be taken as [28],

$$E_g(x) = 1.424 + 1.247x \quad eV \qquad at\ 300K \tag{7.38}$$

where the mole-fraction x is less than 0.4. A more accurate fit may be achieved using the quadratic relationship obtained by Lee et al,

$$E_g(x) = 1.425 + 1.55x + 0.37x^2 \quad eV \qquad at\ 300K \tag{7.39}$$

The compositional dependence of the density of states in the conduction band and valence bands have been described as [28],

$$N_c(x) \simeq 2.5x10^{25}(0.067 + 0.083x)^{\frac{3}{2}} \quad m^{-3} \tag{7.40}$$

$$N_v(x) \simeq 2.5x10^{25}(0.680 + 0.310x)^{\frac{3}{2}} \quad m^{-3} \tag{7.41}$$

The variation in electron affinity is given by,

$$\chi(x) = 4.07 - 1.06x \quad eV \tag{7.42}$$

The net recombination rate may be modelled using a simplified Shockley-Hall-Read expression in which the traps are located at band centre and both carrier species are assumed to have the same lifetime ($\tau = 1$). Experimental evidence supports this simple model.

$$G = \frac{np - n_i^2}{\tau(n + p + 2n_i)} \tag{7.43}$$

Since holes take a less prominent role in the operation of the HBT the hole mobility is taken to be independent of the mole-fraction of aluminium in the AlGaAs. The bulk mobility-field dependence for the holes and electrons may be described by the following relationships [30,31],

$$\mu_p(E) = \frac{\mu_{po}}{1 + \dfrac{\mu_{po}E}{v_{sat}}} \quad m^2V^{-1}s^{-1} \tag{7.44}$$

$$\mu_n(E) = \frac{300\mu no}{T}\left[\frac{1 + \dfrac{8.5\times10^4 E^3}{\mu_{no}E_c^4(1 - 5.3\times10^{-4}T)}}{1 + \left(\dfrac{E}{E_c}\right)^4}\right] \quad m^2V^{-1}s^{-1} \tag{7.45}$$

where V_{sat} is the hole saturation velocity in GaAs ($1.5\times10^5 ms^{-1}$), E_c is the critical field ($4\times10^5 Vm^{-1}$), T is temperature in Kelvin, μ_{po} and μ_{no} are the low field mobilities. The hole and electron low field mobilities may be expressed as a function of doping and temperature as [32],

$$\mu_{po} = \frac{300}{T}\left[\mu_{p1} + \frac{\mu_{p2}}{1 + \left(\dfrac{N_D + N_A}{N_{ref}}\right)^{\alpha}}\right] \quad m^2V^{-1}s^{-1} \tag{7.46}$$

$$\mu_{no} = \frac{0.8}{1 + \left(\dfrac{N_D + N_A}{10^{23}}\right)^{1/2}} \quad m^2V^{-1}s^{-1} \tag{7.47}$$

where $\mu_{p1} = 0.005m^2V^{-1}s^{-1}$, $\mu_{p2} = 0.033m^2V^{-1}s^{-1}$, $N_{ref} = 3.232\times10^{23}m^{-3}$ and $\alpha = 0.4956$. The mobility in the AlGaAs is reduced as the mole-fraction of aluminium x is increased as a result of alloy scattering. This effect may be included in the model by empirically scaling the low field mobility,

$$\mu(x) = m(x)\mu(x=0) \tag{7.48}$$

where $m(x)$ is a monotonically decreasing function of the mole-fraction x. The theoretical relationship between mobility and mole-fraction is not well documented.

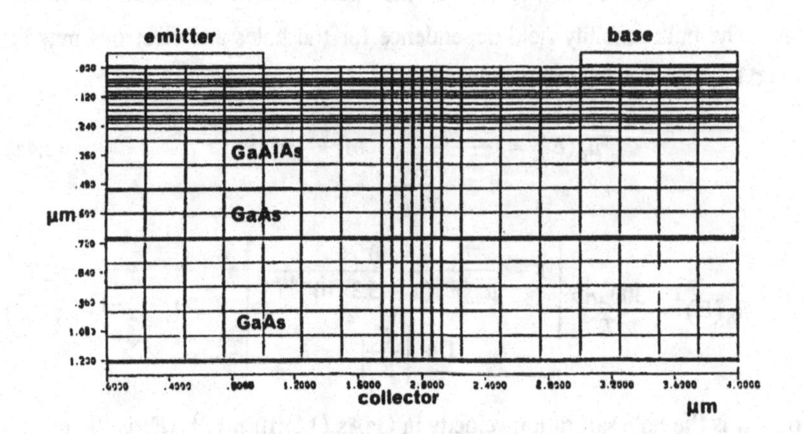

Figure 7.8 Finite-difference mesh for a heterojunction bipolar transistor simulation

A finite-difference mesh used in the simulation of an HBT is shown in Figure 7.8. The cross-sectional area of the transistor modelled in the simulation is 1.2 μm deep by 4.0 μm wide, with 1 μm base and emitter contacts. A variable mesh has been used with closely-packed mesh lines in the vicinity of the junctions, where the potential and carrier density levels vary much more rapidly.

The current continuity equations for the HBT may be solved using finite-difference equations based on the Scharfetter-Gummel algorithm in a similar manner to the homojunction problem. In this case the current density equations are expressed (for the x components only here), for electrons as

$$J_{nx_{i+1/2}} = \frac{D_{n_{i+1/2}}}{a_i}\left\{B\left[\frac{(\psi_i + \theta_{n_i}) - (\psi_{i+1} + \theta_{n_{i+1}})}{V_t}\right]n_i\right.$$

$$\left. B\left[\frac{(\psi_{i+1} + \theta_{n_{i+1}}) - (\psi_i + \theta_{n_i})}{V_t}\right]n_{i+1}\right\} \qquad (7.49)$$

and for holes as

$$J_{px_{i+1/2}} = \frac{D_{p_{i+1/2}}}{a_i}\left\{B\left[\frac{(\psi_{i+1} - \theta_{p_{i+1}}) - (\psi_i - \theta_{p_i})}{V_t}\right]p_i\right.$$

$$\left.B\left[\frac{(\psi_i - \theta_{p_i}) - (\psi_{i+1} - \theta_{p_{i+1}})}{V_t}\right]p_{i+1}\right\} \tag{7.50}$$

where V_T is the thermal voltage (kT/q) and B is the Bernoulli function defined as,

$$B(x) = \frac{x}{\exp(x) - 1} \tag{7.51}$$

The band parameters θ_n and θ_p are defined as

$$\theta_n = \frac{\psi - \psi_0}{q} + \frac{kT}{q} ln\left(\frac{N_C}{N_{C0}}\right) \tag{7.52}$$

and

$$\theta_p = -\frac{\psi - \psi_0}{q} + \frac{kT}{q} ln\left(\frac{N_V}{N_{V0}}\right) - (E_g - E_{go}) \tag{7.53}$$

The Dirichlet boundary conditions for the ohmic contacts can be obtained by setting the quasi-Fermi levels such that equilibrium is imposed at the contacts. Hence the potential boundary conditions may be given as

$$\psi = V_{applied} + \frac{kT}{q} ln\left(\frac{n}{n_i}\right) - \theta_n \tag{7.54}$$

and

$$\psi = V_{applied} - \frac{kT}{q} ln\left(\frac{p}{n_i}\right) + \theta_p \tag{7.55}$$

The corresponding boundary conditions for the carrier densities at the ohmic contacts can be described by,

$$n = \frac{(C^2 + 4n_i^2)^{1/2} + C}{2} \tag{7.56}$$

and

$$p = \frac{(C^2 + 4n_i^2)^{1/2} - C}{2} \tag{7.57}$$

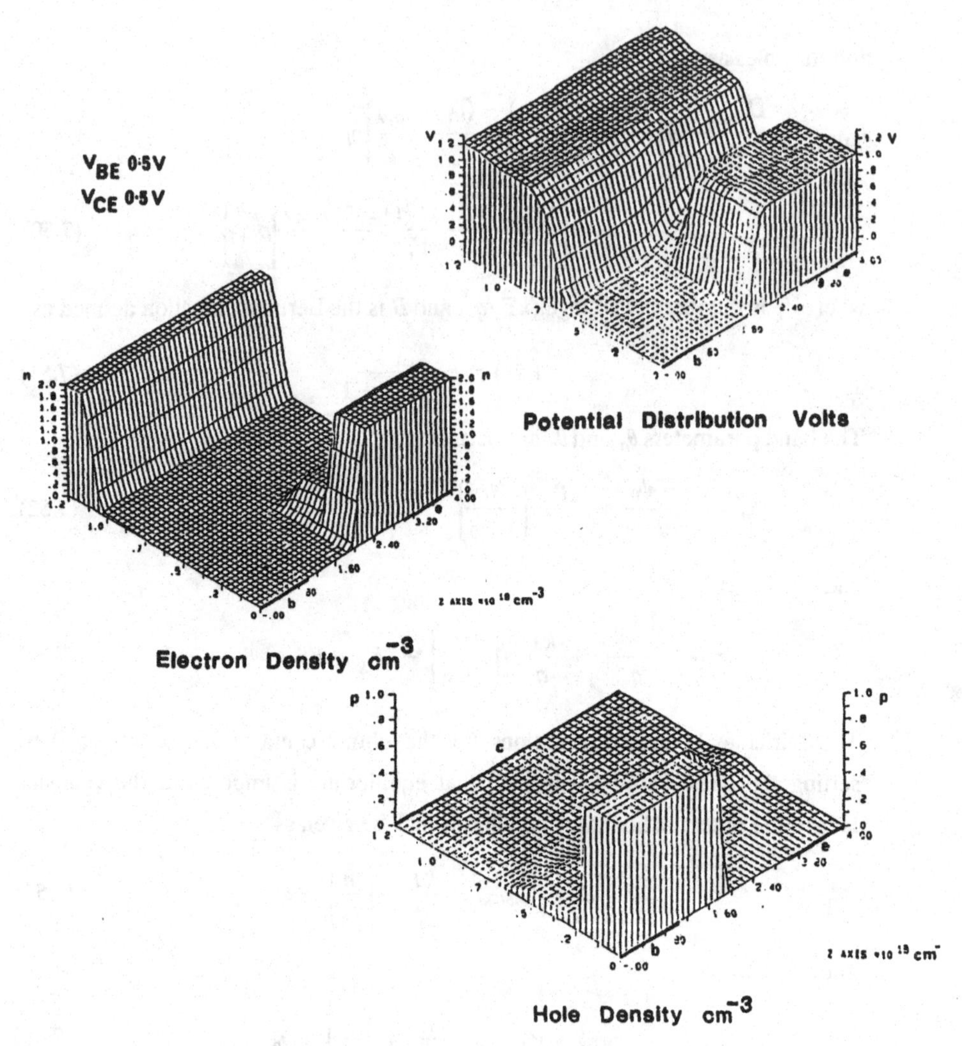

Figure 7.9 Potential, electron and hole density distributions obtained from the HBT simulation

where C accounts for charge defects and donor/acceptor densities (which can in its simplest form become $N_D - N_A$).

The potential, electron and hole distributions for the HBT model shown in Figure 7.8 at a bias of $V_{BE} = 0.5V$ and $V_{CE} = 0.5V$ are shown in Figure 7.9. It can be seen that there is a barrier in the base region corresponding to the p-P heterojunction.

7.4. Monte Carlo Simulations

Monte Carlo methods, discussed in the next chapter, have also been recently applied to the study of heterojunctions [29,33,34]. This type of models allows a detailed treatment of scattering and quantisation effects. The Monte Carlo method provides an important insight into the electron propagation in these devices. The formulation of a quantum mechanical model used in Monte Carlo simulations is discussed in Chapter 9.

References

[1] Anderson, R.L., "Germanium-gallium-arsenide heterojunctions", IBM J.Res. and Develop., Vol.4, pp.283-287, 1960

[2] Delagebeaudeuf, D. and Linh, N.T., "Metal-(n) AlGaAs-GaAs two-dimensional electron gas FET", IEEE Trans. Electron Devices, ED-29, pp.955-960, 1982

[3] Asbeck, P.M., Miller, D.L., Asatourian, R. and Kirkpatrick, C.G., "Numerical simulation of GaAs/GaAlAs heterojunction bipolar transistors", IEEE Electron Device Lett., Vol. EDL-3, pp.403-406, 1982

[4] Lundstrom, M. and Schuelke, R.J., "Modeling semiconductor heterojunctions in equilibrium", Solid State Electron., pp.683-691, 1982

[5] Lundstrom, M.S. and Scheulke, R.J., "Numerical analysis of heterostructure semiconductor devices", IEEE Trans. Electron Devices, ED-30, pp.1151-1158, 1983

[6] Mawby, P.A., Snowden, C.M. and Morgan, D.V., "Numerical modelling of GaAs/AlGaAs heterojunctions and GaAs/AlGaAs heterojunction bipolar transistors", Proc. 2nd Int. Conf. on Simulation of Semiconductor Devices and Processes, Swansea: Pineridge Press, pp.82-99, 1986

[7] Mimura, T. et al, "A new field effect transistor with selectively doped $GaAs/n-Al_xGa_{1-x}As$ heterojunctions", Japan J. Appl. Phys., Vol.19, pp.225-227, 1980

[8] Hiyamizu, S. et al, "High mobility of two-dimensional electrons at the GaAs/n-AlGaAs heterojunction interface", Appl. Phys. Lett., Vol.37, pp.805-807, 1980

[9] Esaki, L. and Tsu, R., "Superlattice and negative conductivity in semiconductors", IBM Res. Report. RC 2418, March 26, 1969

[10] Dingle, R.et al, "Electron mobilities in modulation-doped semiconductor heterojunction superlattices", Appl. Phys. Lett., Vol. 33, pp.665-667, Oct. 1978

[11] Mimura, T et al, "New device structure for 4kb HEMT SRAM", Tech. Dig. IEEE Gallium Arsenide Integrated Circuit Symp., October 1984

[12] Berenz, J., Nakono, K and Weller, K., "Low noise high electron mobility transistors", Tech. Dig. IEEE Microwave and Millimeter Wave Circuits Symposium, San Francisco, pp.83-86, 1984

[13] Drummond, T.J., Morkoc, H., Lee, K. and Shur, M., "Model for modulation doped field effect transistor", IEEE Electron Dev. Lett., EDL-3, No.11, pp.338-341, 1982

[14] Lee, K., Shur, M.S., Drummond, T. and Morkoc, H., "Current-voltage and capacitance-volatage characteristics of modulation-doped field effect transistors", IEEE Trans. Electron Devices, ED-30, pp.207-212, 1983

[15] Lee, K., Shur, M.S., Drummond, T. and Morkoc, H., "Parasitic MESFET in (Al,Ga)As/GaAs modulation doped FET's and MODFET characterisation", IEEE Trans. Electron Devices, ED-31, pp.29-35, 1984

[16] Hughes, W.A., Snowden, C.M., "Non-linear charge control in AlGaAs/GaAs modulation doped FET's", to be published, IEEE Trans. Electron Devices, 1987.

[17] Lee, K., Shur, M.S., Drummond, T.J., Su, S.L., Lyons, W.G., Fischer, R. and Morkoc, H., "Design and fabrication of high transconductance modulation doped (Al,Ga)As/GaAs FETs", J.Vac.Sci.Technol. B, Vol. 1, No.2, pp.186-189, 1983

[18] Yoshida, J. and Kurata, M., "Analysis of high electron mobility transistors based on two-dimensional numerical model", IEEE Electron Device Letters, Vol. EDL-5, pp.508-510, 1984

[19] Tang, J.Y.F., "Two-dimensional simulation of MODFET and GaAs gate heterojunction FET's", IEEE Trans. Electron Devices, ED-32, pp.1817-1823, 1985

[20] Widiger, D., Kizilyalli, I.C., Hess, K. and Coleman, J.J., "Two-dimensional transient simulation of an idealised high electron mobility transistor", IEEE Trans. Electron Devices, Vol. ED-32, pp.1092-1102, 1985

[21] Loret, D., Baets, R., Snowden, C.M. and Hughes, W.A., "Two-dimensional numerical models for the high electron mobility transistor", Proc. 2nd Int. Conf. on Simulation of Semiconductor Devices and Processes, Swansea: Pineridge Press, pp.100-113, 1986

[22] Al-Mudares, M.A.R., "Computer simulation studies of microwave MESFET's", PhD Thesis, University of Surrey, 1984

[23] Schuelke, R.J. and Lundstrom, M.S., "Thermionic emission-diffusion theory of isotype heterojunctions", Solid State Electron., Vol.27., No. 12, pp.1111-1116, 1984

[24] Grinberg, A.A., Shur, M.S., Fisher, R.J. and Morkoc, H., "An investigation of the effect of graded layers and tunneling on the performance of AlGaAs/GaAs heterjunction bipolar transistors", IEEE Trans. Electron Devices, Vol.ED-31, pp.1758-1765, 1984

[25] Baccarani, G., Jacaboni, C. and Mazzone, A.M., "Current transport in narrow base transistors", Solid State Electron., Vol.20, pp.5-10, 1977

[26] Dingle, R., Wiegmann, W. and Henry, C.H., "Quantum states of confined carriers in very thin $Al_xGa_{1-x}As-GaAs-Al_xGa_{1-x}As$ heterostructures", Phys. Rev. Lett., Vol. 33 , pp.827-830, 1974

[27] Kroemer, H., Chien, W. -Y, Harris, J.S. and Edwall, D.D., "Measurement of isotype heterojunction barriers by C-V profiling", Appl. Phys. Lett., Vol. 36, No.4, pp.295-297, 1980.

[28] Casey, H.C. and Panish, M.B., Heterostructure Lasers, New York: Academic Press, 1978

[29] Maziar, C.M, Klausmeier-Brown, M.E., Bandyopadhyay, S., Lundstrom, M.S. and Datta, S., "Monte Carlo evaluation of electron transport in heterojunction bipolar transistor base strucrures", IEEE Trans. Electron Devices, ED-33, pp.881-888, 1986

[30] Snowden, C.M., Howes, M.J. and Morgan, D.V., "Large-signal modelling of GaAs MESFET operation", IEEE Trans. Electron Devices, ED-30, pp.1817-1824, 1983

[31] Selberherr, S., Analysis and simulation of Semiconductor Devices, Wien, New York: Springer-Verlag, pp.80-102, 1984

[32] Hilsum, C., "Simple empirical relationship between mobility and carrier concentration", Electronics Letters, Vol.10, No.12, p.259, 1974

[33] Cahay, M., McLennan, M., Bandyopadhyay, S., Datta, S. and Lundstrom, M.S., "Self-consistent treatment of electron propagation in devices", Proc. 2nd Int. Conf. on Simulation of Semiconductor Devices and Processes, Swansea: Pineridge Press, pp.58-81, 1986

[34] Al-Mudares, M.A.R. and Ridley, B.K., "Monte Carlo scattering-induced negative differential resistance in AlGaAs/GaAs quantum wells", J.Phys. C., pp.3179-3192, 1986

CHAPTER 8

THE MONTE CARLO METHOD

The Monte Carlo method is a statistical numerical technique for solving mathematical and physical problems. It has been used to solve to a wide variety of problems ranging in scale from the microscopic motion of electrons to the macroscopic movements of galaxies. This method was first applied to the modelling of semiconductors in the 1960's. Kurosawa first reported the use of this method for modelling carrier transport in 1966 [1], and interest increased rapidly. During the following three years Rees introduced a self-scattering scheme [2,3], Fawcett and Rees included distribution anisotropy and Fawcett et al investigated nonparabolicity effects [4]. Key developments in the period 1971 to 1977 included the introduction of many-particle simulations [5], harmonic time variation [6], diffusion [7], spatial and temporal transient phenomena [8,9], and the modelling of alloy semiconductors [10]. Monte Carlo techniques have also been applied to the developing area of quantum transport [11].

Monte Carlo techniques have been applied to a wide variety of semiconductor problems including material properties and devices. Particular attention has been paid to the study of mobility and diffusion properties. The method has been used to study the efficiency of scattering mechanisms, energy distributions, valley populations, velocity and effective mass variations. Large-scale Monte Carlo simulations have been applied to a variety of devices including pn junctions, MESFET's and more recently photodiodes [12,13,14].

The Monte Carlo method allows the Boltzmann transport equation to be solved using a statistical numerical approach, by following the transport history of one or more carriers (particles), subject to the action of external forces. The forces on the particles consist of applied electric and magnetic fields, and scattering mechanisms. The essence of the Monte Carlo method lies in the generation of sequences of random numbers with specified distribution probabilities used to describe the microscopic processes such as scattering events which determine the time between successive collisions of carriers. Monte Carlo simulations have been used to model steady-state phenomena in homogeneous samples by simulating the motion of a single electron. This assumes that if the motion of the particle is followed for a sufficiently long period of time, it will yield information on the behaviour of the entire electron gas. An important feature of this approach is that because only one particle is involved it means that a relatively long history can be accumulated in reasonable amounts of computer time. Alternatively, if the sample is not homogeneous or transient transport conditions are of interest, it is necessary to simulate the history of large numbers of carriers.

The Monte Carlo method applied to semiconductors has reached a high level of development. A particularly attractive feature of the method is that it allows the simulation of physical systems which cannot be achieved using laboratory experiments and in some contexts it is viewed as being closely analogous to experimental techniques. It is not possible in the space of one chapter to do more than introduce the reader to the method and further reference may be made to detailed reviews of Monte Carlo semiconductor modelling written by Alberigi-Quaranta et al [15], Price [16], Jacoboni and Reggiani [17,18], Hockney [19] and Moglestue [20].

8.1. The Monte Carlo Method Applied to Carrier Transport in Semiconductors

The Monte Carlo method applied to semiconductor device modelling usually follows the procedures originally described by Kurosawa [1] and Fawcett et al [21]. This involves following the motion of electrons in momentum space, where random numbers control the stochastic scattering process. The Monte Carlo method models the motion of each electron as a sequence of free flights between scattering events (collisions). It is usually assumed that between collisions the electron is accelerated in a constant field and obeys the classical laws of motion governed by the energy band structure of the material.

The exact choice of Monte Carlo simulation scheme is largely governed by the requirement to characterise steady-state or transient phenomena and whether or not the sample in question is homogeneous. This will lead to either a single particle simulation or one requiring an ensemble of particles. Steady-state, homogeneous transport may be modelled using a single-particle simulation, whereas time- and space-dependent phenomena and non-homogeneous systems generally require the motion of a large number of particles to be simulated. The situation is simplified for time- and space-dependent systems which are periodic in space or time and can be modelled using a single particle simulation.

A Monte Carlo simulation requires a detailed definition of the physical system as a starting point. In the case of electron transport, which is considered in the following sections, this requires material parameters, a knowledge of the energy band structure, lattice temperature, definition of the applied external fields and a set of initial conditions. The electron has three momentum components associated with it, usually determined from a Maxwellian distribution at a particular lattice temperature. The velocity components associated with the distribution function are selected using a random number generator.

In the case of steady-state simulations the duration of the simulation must be long enough (usually greater than 20,000 interactions) so that the choice of initial conditions does not influence the final results. This can be achieved by dividing the simulation into several 'subhistories' and using the final state of previous subhistories as the initial state of each new subhistory. Transient simulations are not carried out in this way since the initial distribution of electron states for a specific physical system determines the initial transient and is an integral part of the process which determines the final results.

The process of simulating the particle motion involves a number of computational steps to determine the duration of free flight, the scattering mechanism and the choice of state after scattering. The data collected from each free flight is then used to determine the required parameters and its eact treatment will depend on the nature of the simulation (steady-state or transient).

8.1.1. Equations of Motion, Energy Band Structure and Free Flight

The electric field **E**, which forms a driving force for the simulation, is obtained from potential distribution ψ by solving Poisson's equation.

$$\nabla \mathbf{E} = -\nabla^2 \psi = \frac{1}{\varepsilon_r \varepsilon_0} \rho \tag{8.1}$$

where ρ is the charge density due to ionised dopants and charge carriers. The solution of the Poisson equation may be carried out using a conventional finite-difference mesh superimposed on the simulation domain. In the case of a two-dimensional model, a cross-sectional mesh is superimposed over the device geometry, dividing the device into cells with node spacings Δx and Δy.

The charge due to ionized dopants and charge carriers is assigned to each cell using a suitable technique (for example [22][23]). A particularly popular method is the Nearest Grid Point method, where the charge in each cell is taken to act at its midpoint. In very small devices (dimensions of less than 0.1 μm), or at high doping densities (greater than $10^{24} m^{-3}$, other more accurate methods must be employed [23,24]. The total number of charge carriers in a semiconductor sample of volume V with a uniform density of ionised impurities N is NV. For example a modern microwave GaAs MESFET could have of the order of 10^7 carriers in the active channel. This number of charge carriers is far too large to be analysed on a computer and it is necessary to restrict simulations to less than 10^5 carriers. In order to satisfy the Poisson equation it is necessary to assign a charge q_p to each particle where

$$n_c q_p = q \, \Delta x \, \Delta y \, \Delta z \, N_o \tag{8.2}$$

where Δz is the dimension of the cell perpendicular to the x-y plane and n_c is the number of modelled particles per cell. These particles of charge q_p are referred to as super particles.

Poisson's equation is solved using direct, iterative or fast fourier transform (FFT) techniques. A popular choice is Hockney's POT4 FFT scheme [25]. The electric field **E** may be calculated using a standard finite-difference approximation. The electric field **E** is assumed to be constant within each mesh cell, but may change from cell to cell. The calculated value of the electric field is also assumed to be constant between time steps. The time step Δt is chosen to be as long as possible to minimise computation time, but short enough to minimise errors due to the discretized nature of the electric field and is usually chosen to lie in the range in the range 2 to 50 fs.

The equations of motion governing the electron during free flight are

$$\hbar \frac{d\mathbf{k}}{dt} = q\mathbf{E} \tag{8.3}$$

where $\hbar$ is the reduced Planck's Constant

$$\hbar = \frac{h}{2\pi} = 1.054 \times 10^{-34} Js \tag{8.4}$$

$$\mathbf{v} = \frac{d\mathbf{r}}{dt} = \frac{1}{\hbar} \frac{\partial \xi(\mathbf{k})}{\partial \mathbf{k}} \tag{8.5}$$

where $\mathbf{v}$ is the particle velocity $\mathbf{r}$ the position, $\mathbf{k}$ is the electron wave vector, ξ the electron energy, and q is the charge on the electron (which is negative). Equation (8.3) is also known as the quasiclassical approximation [7]. The electric field, which is assumed to be constant during free flight, changes the electron wavevector $\mathbf{k}$ linearly with time during free flight.

The electron energy in equation (8.5) may be defined in terms of the energy band structure. A three valley model of the energy band structure is usually assumed, with the Γ (<000> direction), X (<100> direction) and L (<111> direction) bands, centred in the first Brillouin zone, Figure 8.1. The electron and hole energies, referenced to the bottom of the conduction band or top of the valence band respectively, can be obtained from a variety of relationships. These are generally based on a quadratic function of $\mathbf{k}$, describing a parabolic band structure. The simplest band structure assumes spherically equienergetic surfaces with a single scalar effective mass and is described by

$$\xi(\mathbf{k}) = \frac{\hbar k^2}{2m^*} \tag{8.6}$$

where m^* is the effective mass and k is the magnitude of $\mathbf{k}$. This model is appropriate for modelling the minimum of the conduction band located at Γ and for the maximum of the valence band. Because of its simplicity it is often used to obtain approximate transport parameters where accuracy is not essential.

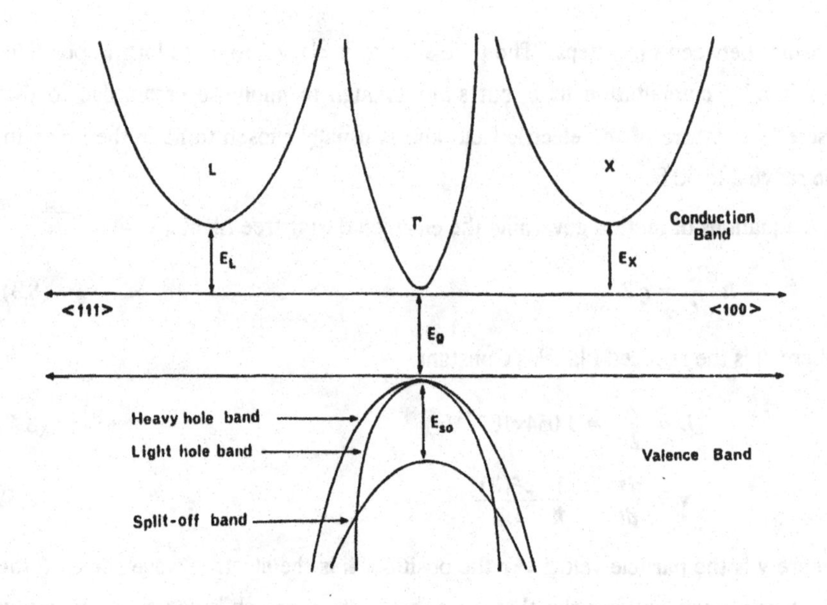

Figure 8.1 Band structure of a cubic semiconductor

The electron energy for cubic semiconductors with ellipsoidal equienergetic surfaces, may be taken as

$$\xi(\mathbf{k}) = \frac{1}{2}\hbar^2\left(\frac{k_x^2}{m_x{}^*} + \frac{k_y^2}{m_y{}^*} + \frac{k_z^2}{m_z{}^*}\right) \tag{8.7}$$

where k_x, k_y and k_z are the k vectors and $m_x{}^*$, $m_y{}^*$ and $m_z{}^*$ are the effective masses along the three Cartesian axes (tensor effective mass). In GaAs and InP the three effective masses are all equal (m^*).

For a band structure with warped equienergetic surfaces, such as in the case of the two degenerate maxima in the valence band, other functions are available [26].

At values of **k** far from the conduction band minima and valence band maxima the energy may not follow a parabolic function. In this situation of non-parabolicity the conduction band energy-wave-vector functions can be modified by replacing $\xi(\mathbf{k})$ by $\xi(1 + \alpha\xi)$ in the above expressions. For example in the simplest case,

$$\xi(1 + \alpha\xi) = \gamma(\mathbf{k}) = \frac{\hbar k^2}{2m^*} \tag{8.8}$$

or alternatively,

$$\xi(\mathbf{k}) = \frac{-1 + (1 + 4\alpha\gamma)}{2\alpha} \tag{8.9}$$

where α is a non-parabolicity factor and for the case of GaAs is zero except for the Γ conduction band. In the case of silicon the effect of band non-parabolicity is significant at high values of electric field, where the electrons have reached energies far from the bottom of the band. Although non-parabolicity is apparent for lower energies in germanium it is it is less significant at higher energies and is often neglected. Non-parabolicity in the valence band cannot be treated in the same way because its effect depends on the direction and energy level. At very high energy levels (greater than 1 eV) the simple band structure described above may not adequately describe the electron energy.

The velocity, energy and position of each particle change continuously during free-flight and are determined from equations (8.3) to (8.8) prior to scattering.

If the particle was last scattered at $t = 0$, then the probability of the particle being scattered between time t and $t + dt$ is determined by,

$$P_f(t)dt = \lambda\left[\xi(t)\right] \exp\left\{-\int_0^t \lambda\left[\xi(t')\right] dt'\right\} dt \tag{8.10}$$

where $P_f(t)$ is the probability density of scattering at time t and $\lambda(\xi)$ is the total scattering rate determined from scattering mechanism calculations. The free flight time $t = t_f$ is obtained by generating random numbers with the probability distribution in equation (8.10), according to the relationship,

$$P_f(t)dt = P(r_f)dr_f \tag{8.11}$$

where the random number r_f lies in the range 0 to 1 and

$$\int_0^t P_f(t)dt = \int_0^{r_f} P(r_f')dr_f' \tag{8.12}$$

Hence

$$r_f = 1 - \exp\left(-\int_0^{t_f} \lambda(t)dt\right) \tag{8.13}$$

which can alternatively be written

$$\int_0^{t_f} \lambda(t)dt = - ln(1 - r_f) \tag{8.14}$$

The time of flight t_f is obtained by generating a random number r_f and numerically integrating equation (8.14).

The above method for determing the time of flight t_f requires the solution of an integral equation for each scattering event. As an alternative this Rees [2,3] introduced a notional 'self-scattering' concept to overcome difficulties in solving equation (8.10). In circumstances where the random selection process for scattering does not select one of the 'physical' scattering mechanisms outlined in the following section, self-scattering is imposed. Using this approximation, the probability density of scattering in equation (8.10) reduces to

$$P_f(t) = \lambda_{ss} \exp\left(-\lambda_{ss}t\right) \tag{8.15}$$

where λ_{ss} is the Rees self-scattering rate.

$$\lambda_{ss} = \sum_{i=0}^{n} \lambda[k(t)] \tag{8.16}$$

A random number r_f with a uniform density is used to determine the time of free flight t_f

$$t_f = - \frac{1}{\lambda_{ss}} ln(r) \tag{8.17}$$

The parameter λ_{ss} is chosen so that it is constant, greater than zero and not less than the sum of all the scattering rates λ over the full range of electron energies. If the particle undergoes self-scattering, it state $|\mathbf{k}'\rangle$ after the collision is taken to be equal to the state $|\mathbf{k}\rangle$ before the collision. This means that the particle will continue unperturbed with the same momentum and energy as it had before the collison. If self-scattering predominates there will be several positions of the free flight trajectory where the electron will not experience any change in momentum or energy. In order to save on computation time it is useful to reduce self-scattering to a minimum. In order to avoid excessive computation time a simple iterative scheme is usually used to determine λ_{ss}.

After a free flight of duration t the position $\mathbf{r}$ and momentum $\mathbf{k}$ are given by,

$$\mathbf{r} = \mathbf{r}_o + \frac{\hbar \mathbf{k}_o t}{m^*} + \frac{q}{m^*}\int_0^t dt' \int_0^{t'} dt'' \mathbf{E}(t'') \quad (8.18)$$

$$\mathbf{k} = \mathbf{k}_o + \frac{q}{\hbar}\int_0^t dt' \mathbf{E}(t') \quad (8.19)$$

$\mathbf{r}_o$ and $\mathbf{k}_o$ are the position and wave vector of the particle immediately after the last scattering event. When particles reach the boundary of the device they are reflected or absorbed depending on the nature of the boundary conditions.

Increased computational efficiency can be achieved by allowing the particles which cross cell boundaries to continue to accelerate as a function of the electric field they experienced before crossing the boundary. Under this assumption the particle position and momentum are defined as,

$$\mathbf{r} = \mathbf{r}_o + \frac{\hbar \mathbf{k}_o t}{m^*} + \frac{q}{2m^*} \mathbf{E} t_f^2 \quad (8.20)$$

$$\mathbf{k} = \mathbf{k}_o + \frac{q \mathbf{E} t_f}{\hbar} \quad (8.21)$$

The particle then continues to accelerate until the next scattering event or the end of the time step (sometimes referred to as the field-adjusting time step). If the particle has remained in free flight at the end of the time step, it will continue to accelerate under the influence of the new local electric field for the remainder of its free flight during the next time step. In order to ensure that the time step is short enough so that only relatively few particles cross the cell boundaries, the time step may be chosen according to the relationship

$$\Delta t \simeq \frac{\min(\Delta x, \Delta y)}{v} \quad (8.22)$$

where v is the average velocity of the particles. Stochastic heating effects and numerical stability requirements further restrict the size of the time step Δt [27].

8.1.2. Scattering Mechanisms

At the end of the free flight, the electron is scattered by a scattering mechanism chosen according to the relative propability of all possible scattering mechanisms. The occurence of the various scattering mechanisms also depends on physical parameters such as energy, temperature, electric and magnetic field strengths. Scattering mechanisms which operate on the electron include acoustic and optical phonon scattering, intervalley scattering, ionised impurity (Coulomb) scattering, piezoelectric scattering, alloy scattering and neutral ionised scattering. The latter effect is only significant at very low temperatures and is usually neglected. Since optical phonon dispersion is reasonably flat it is usual to consider only one phonon energy. Acoustic phonon scattering is often regarded as elastic because the acoustic phonon energy is small relative to the electron energy. Impurity scattering is assumed to be elastic and has a strong dependence on wave vector **k** [20,28]. Scattering mechanisms are usually assumed to be an instantaneous process although Barker and Ferry have demonstrated that for very small devices ($< 0.1\ \mu m$) the interaction times must be accounted for [29].

When an electron is scattered its wavevector changes from **k** to **k**′. The scattering process is classified in terms of the total scattering rate and scattering angles. This requires a knowledge of the transitions of an electron between the different states **k** and **k**′. In order to simplify the analysis here, it will be assumed that the collison duration τ_c is zero. This approximation is reasonable for most applications but may not be valid for very small scale devices [29]. The transition probability per unit time $P(\mathbf{k},\mathbf{k}')$ from state $|\mathbf{k}>$ to $|\mathbf{k}>'$ is given to a first-order by Fermi's golden rule,

$$P(\mathbf{k},\mathbf{k}') = \frac{2\pi}{\hbar}\,|<\mathbf{k}'|H'|\mathbf{k}>|^2\delta[\xi(\mathbf{k}') - \xi(\mathbf{k}) \pm \Delta xi] \tag{8.23}$$

where H' is a perturbating Hamiltonian (scattering potential) and $<\mathbf{k}'|H'|\mathbf{k}>$ is the matrix element of the Hamiltonian between the initial and final states. $\Delta\xi$ is the energy change (positive or negative) during the transition. The total scattering rate out of state $|\mathbf{k}>$ may be obtained from the summation of equation (8.23) over all allowed final states,

$$\frac{1}{\tau(\mathbf{k})} = \sum_{\mathbf{k}'} P(\mathbf{k},\mathbf{k}') \tag{8.24}$$

where $1/\tau(\mathbf{k})$ is the total scattering rate. If this summation is approximated as an integral and then expressed in spherical coordinates the total scattering rate may be written as,

$$\frac{1}{\tau(\mathbf{k})} = \frac{V}{(2\pi)^3}\int_{\mathbf{k}'} P(\mathbf{k},\mathbf{k}')\,d\mathbf{k}'$$

$$= \frac{V}{8\pi^3}\int_0^{\infty}\int_{-\pi/2}^{\pi/2}\int_0^{\pi} P(\mathbf{k},\mathbf{k}')k'^2\cos\theta d\phi d\theta dk' \tag{8.25}$$

where V is the volume of the crystal, ϕ is the azimuth angle and θ represents the angle between $\mathbf{k}$ and $\mathbf{k}'$.

A detailed summary of scattering processes incorporated into Monte Carlo simulations is given in the review of Jacoboni and Reggiani [18].

The total scattering rate is usually expressed as a function of electron energy. This can be achieved by substituting for the energy for ξ (such as equation (8.8)) in eqaution (8.25). The total scattering rate $\lambda(\xi)$ is given by the sum of the individual scattering rates over all the scattering processes,

$$\lambda(\xi) = \sum_{i=1} \frac{1}{\tau_i(\xi)} \tag{8.26}$$

where the scattering rate for a particular scattering mechanism (acoustic, optical phonon etc) is $1/\tau_i(\xi)$ determined from the energy dependent form of equation (8.25).

The angular probability density $P(\theta)$ which describes the angular dependence of the scattering is related to the transition rate $P(\mathbf{k},\mathbf{k}')$ by

$$P(\theta)d\theta \propto P(\mathbf{k},\mathbf{k}')\cos\theta d\theta \tag{8.27}$$

where θ is the angle between $\mathbf{k}$ and $\mathbf{k}'$. The constant of proportionality is obtained by normalising $P(\theta)$,

$$\int_{-\pi/2}^{\pi/2} P(\theta)\,d\theta = 1 \tag{8.28}$$

The two scattering angles ϕ and θ are determined by generating random numbers r_ϕ and r_θ respectively, where the azimuth angle ϕ is given by

$$\phi = 2\pi r_\phi \tag{8.29}$$

and θ is determined according to the relationship

$$r_\theta = \int_{-\pi/2}^{\theta} P(\theta')d\theta' \tag{8.30}$$

The random numbers r_ϕ and r_θ lie in the range 0 to 1.

The scattering mechanism is chosen randomly. The relative probability of the *i'th* scattering mechanism is

$$P_i(\xi) = \frac{\lambda_i(\xi)}{\lambda(\xi)} \tag{8.31}$$

The relative probabilities of each scattering mechanism are stored in a look-up table by defining functions for each scattering mechanism. The scattering mechanism is then selected using a random number generator .

8.2. Treatment of Results

The approach used to determine the parameters of interest from results obtained using the Monte Carlo process depends on the nature of the problem - whether it is a steady-state phenomenon or time- and space-dependent. In the case of a steady-state simulation, where a single particle is considered, the time averaged value of a quantity a[**k**(t)] obtained during a single history over a period T is

$$<a>_T = \frac{1}{T}\int_0^T a\big[\mathbf{k}(t)\big]\,dt = \frac{1}{T}\sum_i \int_0^{t_i} a\big[\mathbf{k}(t')\big]dt' \tag{8.32}$$

The integral over the total simulation time T is separated into the sum of integrals over all flights of duration t_i. Using this expression it is possible to extract time averaged values of drift velocity, average carrier energy and carrier density. The electron distribution function, which is the solution to the Boltzmann transport equation, can also be extracted in a similar way [21] by calculating the normalised time spent by the electron in each cell of a **k** space mesh, evaluated over a long period T.

An alternative method for obtaining a time average in steady-state simulations using a synchronous-ensemble technique has been described by Price [30,31]. The average value of a quantity a is defined as the ensemble average at time t over the N particles of the system using the relationship,

$$<a> = \frac{1}{N} \sum_i a_i \; (t_i = t) \tag{8.33}$$

It can be shown that a steady-state average suitable for Monte Carlo simulations which covers all N electron free-flights, takes the form [18],

$$<a> = \frac{1}{N} \sum_i a_{ie} \tag{8.34}$$

where a_{ie} is the value of the quantity a at the end of the free flight immediately prior to the *i'th* scattering event. Special care must be taken when using this relationship with self-scattering which has a step-shaped scattering probability.

An example of the steady-state evaluation of average momentum and Kinetic energy is given by Aas and Blotekjaer [32] where the averages are calculated for approximately 100 collisions. The calculation is repeated for several such time intervals and is terminated when the deviation of average momentum and kinetic energy are both less than one or two percent, or if 40,000 collisions have been exceeded.

The statistical uncertainty associated with time-averaged results due to the finite simulation time T can be estimated by dividing the history into N sub-histories of duration T/N. If the sub-histories are sufficiently long to be considered independent, but short enough to allow a large number N of them to be obtained, the uncertainty decreases in a $1/\sqrt{N}$ manner.

Time- and space-dependent phenomena require multi-particle simulations with ensemble averages based on equation (8.33) to obtain the required quantities, and it is not possible to use the steady-state relations (8.32) and (8.34). The number of simulated particles must be sufficiently large so that the the ensemble average value of the quantity of interest obtained from equation (8.33) as a function of time will be representative of the average for the entire gas.

$$<a(t)> = \frac{1}{N} \sum_i a_i(t) \tag{8.35}$$

where i applies to all N particles.

The transient response obtained from the simulation depends on the initial conditions of the carriers which should be related to the problem. A Maxwellian distribution at the lattice temperature and no drift velocity, or a Maxwellian distribution with an electron temperature higher than that of the lattice are suitable initial conditions for this type of model.

Space-dependent phenomena, where the carrier transport properties depend upon the spatial distribution of the carriers in the device, also require an ensemble of independent particles to be used, with averages taken over particles at specific positions.

The accuracy of the results for time- and space-dependent simulations can be estimated by separating the entire ensemble into a number of sub-ensembles. The quantity of interest a is estimated for each sub-ensemble. The statistical uncertainty is then obtained from the standard deviation of these values.

The calculation of the response of charge carriers to periodic external fields has been investigated using Monte Carlo simulations. The average electron velocity, which is a function of the periodic external field, can be expressed as a Fourier series. For small signal levels it is convenient to assume that the average electron velocity contains sine and cosine Fourier coefficients only at the fundamental frequency. Large-signal periodic fields lead to a harmonically rich description of the velocity, which is prone to statistical noise. In these circumstances the response may be obtained without Fourier analysis by sampling the particle velocity at fixed intervals, corresponding to definite phases in the period of the external field. This sampling interval lies typically in the range 5×10^{-15} to 5×10^{-14} seconds. Lebwohl has extended the synchronous-ensemble technique for static fields to include periodic fields [33].

8.3. Application of Monte Carlo Simulations

8.3.1. Transport Characteristics

Monte Carlo simulations are used to obtain material characteristics and for directly simulating device operation. The application of Monte Carlo techniques for determining the dependence of parameters such as mobility, diffusion, energy and effective mass on applied electric field, is essential for the successful implementation of detailed bulk transport models based on the semiconductor equations, discussed earlier in this book. The Monte Carlo method also allows the direct simulation of semiconductor devices which is a particularly significant for small-scale devices where the transport does not necessarily obey the classical semiconductor equations. A limitation of the method in this latter application is that numerical noise and restrictive simulation times may inhibit its use for characterising established devices, at the present time.

Monte Carlo simulations are used to determine the drift velocity in samples of semiconductor material for both steady-state and transient conditions. The simulation of bulk properties can be expedited by keeping the particles stationary to avoid interaction with the boundaries and perturbations in the charge density and electric field. This is achieved by calculating updated momentum and energies at the end of each free flight and neglecting the position of the carrier. A steady-state simulation for GaAs starts off with the electrons in the lower band with a Maxwellian velocity distribution at the lattice temperature. The Monte Carlo simulation allows the average drift velocity to be computed as a function of time, for a given field. It is necessary to allow several picoseconds before the carriers reach their steady state velocity. This is because it takes a finite time for carriers to populate the higher energy bands (valleys). During this transient process the carriers may be accelerated to a peak velocity much higher than the steady state value (velocity overshoot) before settling to the final value. There has been considerable effort devoted to the calculation of velocity-field characteristics [21,34,35,36,37]. The steady-state velocity-field characteristics obtained from various published results are shown in Figure 8.2.

Transient results obtained by Ruch [8] and Maloney and Frey [36] who used a modified constant field technique [21] are shown in Figure 8.1 for Si, GaAs, and InP.

Steady-state carrier mobility characteristics $\mu(E)$ may be readily obtained as a function of electric field from the steady state velocity using the relationship,

$$\mu(E) = \frac{\nu(E)}{E} \tag{8.36}$$

where $\nu(E)$ is the average velocity and E the electric field.

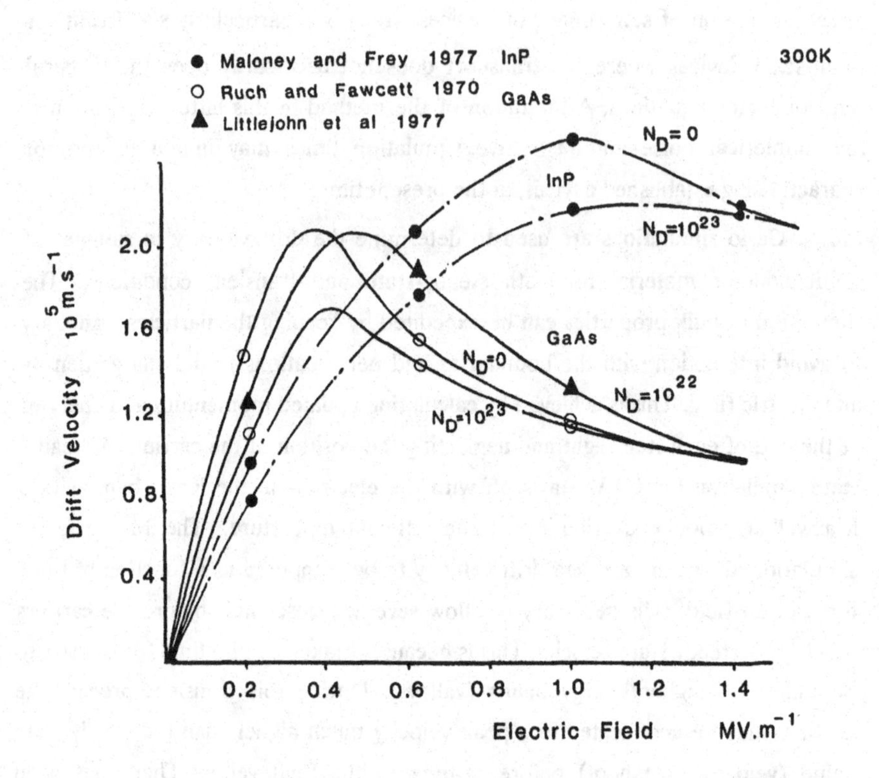

Figure 8.2 Steady-state velocity-field characteristics from Monte Carlo models

Diffusion in semiconductors is a particularly important example of a space- and time-dependent phenomenon, and has been the subject of considerable research using Monte Carlo simulations [4,38,39,40,41]. The diffusion coefficient D can be obtained for low values of electric field, using the Einstein relationship by substituting information on the low field (ohmic) mobility μ,

$$D = \frac{\mu kT}{q} \tag{8.37}$$

However, for high values of electric field, the Einstein relationship no longer holds and it is necessary to consider the influence of hot-electron effects. Intervalley diffusion effects, present in compound semiconductors, also lead to departures from Einstein diffusion even at relatively low field strengths.

If a linear response is assumed, where hot electron and other transient effects are not present, diffusion is often modelled using the relationship (in one-dimension),

$$J_i = qn(\mathbf{x})v_i(\mathbf{E}) - D_{ij}\frac{\partial n(\mathbf{x})}{\partial x_j} \tag{8.38}$$

where J_i is the current density, $\mathbf{x}$ is the position of the particle in space, $n(\mathbf{x})$ the particle density, v is the drift velocity and D_{ij} is the diffusion tensor. This assumes that the drift and diffusion do not influence each other which is not strictly true. Consider a reverse-biased p-n junction or Schottky barrier diode with the diffusion and drift components balancing to produce almost zero current. The energy distribution function does not follow the usual hot electron distribution for a homogeneous system since there is no net exchange of power with the electron gas even though the electrons experience high electric fields between collisions. If the electric field and drift velocity are zero and diffusion still occurs, the equation reduces the first Fick diffusion equation. If the current density expression (8.38) is substituted into the continuity equation, the well known diffusion equation is obtained (also known as the second Fick's equation),

$$\frac{\partial n}{\partial t} = -v_i(\mathbf{E})\frac{\partial n}{\partial x_i} + D_{ij}\frac{\partial^2 n}{\partial x_i \partial x_j} \tag{8.39}$$

A rigourous derivation requires that the drift velocity v and diffusion coefficent D are spatially independent in this equation. At low fields, linear-response conditions are usually assumed and the drift velocity is varies linearly with field and the diffusion coefficient is independent of the field value. At higher fields the diffusion coefficent is usually assumed to be a function of electric field $D(\mathbf{E})$. The diffusion coefficient D or $D(\mathbf{E})$ can be obtained from the solution of the Boltzmann equation. However, field-dependent values $D(\mathbf{E})$ can only be obtained from the Boltzmann solution for small carrier density gradients and by assuming time periods much longer than transient-transport times.

In order to simplify the analysis, the electron density n is assumed to be a function of only one coordinate z, which is parallel to the direction of the drift velocity. The component of diffusion coefficient along the direction of the drift velocity D_l can be calculated from the variance of the distance z travelled by the particles in time t,

$$D_l = \frac{(<z^2> - <z>^2)}{2t} \tag{8.40}$$

This method is frequently used to calculate D_l [7,42,43]. Very long simulation times are required to obtain an accurate estimate of D_l.

An alternative and computationally faster approach is to base the Monte Carlo calculation of the Diffusion coefficient on the second central moment of the carrier distribution function where,

$$D_l = \frac{1}{2}\frac{d}{dt} <(z - <z>)^2> \tag{8.41}$$

In this Monte Carlo calculation a number of particles are independently simulated over a long period of time with their positions are recorded at specific times. The diffusion coefficent D_l is then obtained from the slope of the characteristic. Long simulation times are required to ensure that the second central moment follows a linear dependence predicted by equation (8.41), although this formulation produces an accurate estimate of D_l in less time than methods based on equation (8.40).

Other components of the diffusion can be obtained using an expression analogous to equation (8.41),

$$D_{xy} = \frac{d}{dt} <(x - <x>)(y - <y>)> \tag{8.42}$$

Typical diffusion results obtained from Monte Carlo simulations are shown in Figure 8.3.

Fick's diffusion law cannot be used to describe transient behaviour. It is also important to appreciate that this classical model cannot be applied if the time during which the concentration gradient varies is short compared with energy relaxation times or if the distances at which the gradient varies appreciably is large compared with the electron mean free path. Jacoboni et al [44,45] and others have suggested more generalised diffusion theories which include the effects of diffusion at small distances and short times.

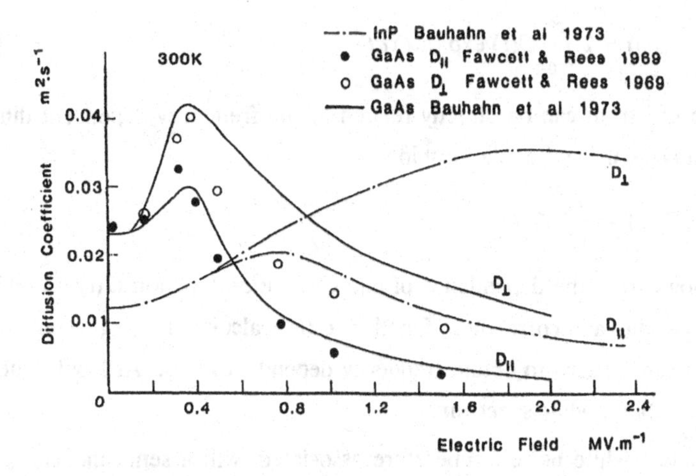

Figure 8.3 Diffusion characteristics obtained from Monte Carlo calculations

Monte Carlo simulations have been used to analyse the noise spectrum of velocity fluctuations in the semiconductor [43,46,47]. The velocity autocorrelation function $C_{ij}(t)$ can be used to determine both the diffusion coefficient D and noise spectrum,

$$C_{ij}(t) = \overline{<\delta v_i(t')\delta v_j(t'+t)>} \tag{8.43}$$

where the bar indicates a time average. $C(t)$ is evaluated by sampling and recording the velocity of a particle at regular intervals over a period longer than the autocorrelation time, and then evaluating the product,

$$v(i\Delta t)v[(i-j)\Delta t] \qquad for\ j=0,1,2...N \tag{8.44}$$

for each i. The time interval Δt is obtained by dividing the total period t into N intervals where $i \leq N$. Products corresponding to the same value of i are averaged over the simulation to obtain

$$C(j\Delta t) + v^2 = \overline{v(t)v(t + j\Delta t} \tag{8.45}$$

In the steady state the ensemble average is included in the time average. The diffusion coefficient D is given by,

$$D = \int_0^\infty C(t)dt \tag{8.46}$$

The noise spectrum $S_v(\omega)$ is obtained using the Wiener-Khintchine theorem,

$$S_v(\omega) = 2\int_0^\infty C(t)\exp(i\omega t)\,dt \qquad (8.47)$$

The noise spectrum can be directly related to the frequency-dependent diffusion coefficient $D(\omega)$ using the relationship

$$D(\omega) = \frac{1}{2}S_v(\omega) \qquad (8.48)$$

which follows from the dependence of autocorrelation function $C(t)$ on diffusion [48]. Hence the autocorrelation function $C(t)$ calculated using Monte Carlo techniques can be used to obtain frequency dependent $D(\omega)$. At low frequencies the noise exhibits a white spectrum.

The equivalent white noise temperature associated with a semiconductor device which exhibits a positive value of the real small-signal impedance, can be extracted using Monte Carlo methods. The field dependent drift velocity $v_d(E)$ and diffusion coefficient $D(E)$ are obtained from the Monte Carlo model and inserted into the expression for noise temperature T_n. For the homogeneous case a lengthy expression for T_n reduces to a form similar to the Einstein relationship [18],

$$T_n = \frac{qD}{k\mu_d} \qquad (8.49)$$

where μ_d is the differential mobility.

Impact ionization in semiconductor devices has been investigated using Monte Carlo simulations by introducing the probability of impact ionization as an independent scattering mechanism in addition to phonon and impurity scattering mechanisms. Ensemble Monte Carlo methods have been used to model impact ionisation, where the electron-hole pair generation rate per particle per unit time g_I is obtained by counting ionization events [49]. The impact ionization rate α_I is then obtained using

$$\alpha_I = \frac{g_I}{v_d} \qquad (8.50)$$

Impact ionization rates have also been estimated using single-particle Monte Carlo simulations, where the impact ionization rate α_I is obtained by averaging the distance to impact ionization over a large number of ionizations [50,51].

A further area of interest applicable to Monte Carlo simulations is in the modelling of semiconductors subject to a combination of electric and magnetic fields. In this situation the equation of motion for a carrier in free flight is modified to

$$\hbar \frac{d\mathbf{k}}{dt} = q\mathbf{E} + \frac{q}{c}\mathbf{v} \times \mathbf{B} \tag{8.51}$$

where $\mathbf{v}$ the group velocity of the particles, $\mathbf{B}$ is the magnetic field and c is the velocity of light. This type of model requires a full three-dimensional simulation since the drift velocity will generally be parallel with the applied electric field and cylindrical symmetry no longer applies. The Hall mobility μ_H, drift mobility μ and longitudinal mobility μ_E are all readily obtained using this type of simulation.

8.3.2. Application of Monte Carlo Techniques to Device Modelling

Monte Carlo simulations provide an important alternative to classical models for investigating the operation of very small semiconductor devices where the carrier transport mechanism is governed by non-stationary processes and classical semiconductor equations no longer apply. Although Monte Carlo models require extensive computer facilities which can restrict the present capabilities of this technique for some applications, they have been used to study a variety of semiconductors device problems. Devices which have been investigated using Monte Carlo simulations include TEDs [5,52], MOSFETs [53], MESFETs [13,52,54,55,56], bipolar transisitors [9], heterostructures [57,58], Schottky diodes [59], and photodetectors [14].

MESFETs are amongst the most extensively researched devices using Monte Carlo methods. The steady-state carrier distribution for single and dual-gate GaAs and Si FETs are shown in Figure 8.4, which clearly reveals the depletion regions beneath each Schottky barrier gate and the non-abrupt nature of the depletion boundaries. The contacts and gaps are all 1 μm in the single gate MESFET, with an epitaxial layer thickness of 0.23 μm and substrate 0.385 μm thick. The epitaxial layer is doped at $N_D = 10^{21}\ m^{-3}$. A gate potential of $-0.1\ V$ and drain potentials of 3 V for the GaAs MESFET and 2V for the Si MESFET are applied. The results are shown at a time of 9.5 ps after the start of the simulation.

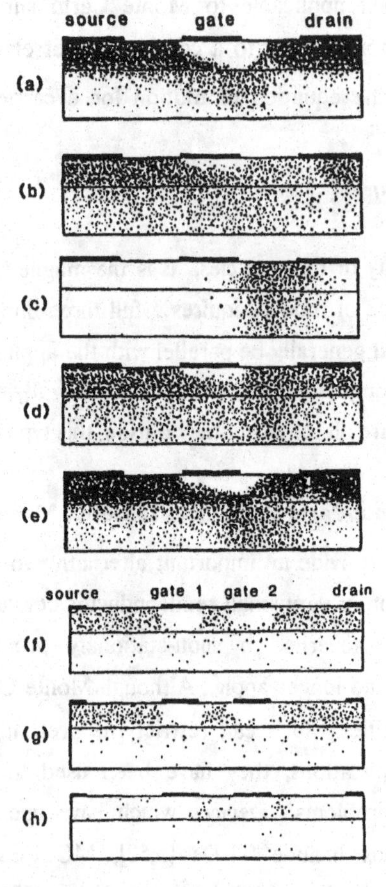

Figure 8.4 Single and dual-gate MESFET Monte Carlo simulation results showing electron density [52,56].

(a) Distribution of central-valley electrons in GaAs MESFET.

(b) Distribution of the upper-valley electrons.

(c) Distribution of all electrons in the microscopic model.

(d) Distribution of all electrons in the diffusive-particle model.

(e) Distribution of all electrons in a Si MESFET.

(f) Distribution of all electrons in a GaAs dual-gate MESFET.

(g) Distribution of Γ electrons.

(h) Distribution of upper-valley electrons.

Moglestue has found that the simulated transconductances for this dual-gate device agree well with measured values corresponding to the real device [56]. The electron density distributions in Figure 8.4 clearly show the penetration of carriers into the substrate decreasing the transconductance g_m and increasing the output conductance. The work of Hockney et al [52] demonstrated that the commonly used diffusion approximation is inadequate for modelling the channel region in GaAs MESFETs and have shown that the assumption of an instantaneous mobility relationship leads to the stationary accumulation region appearing under the drain edge of the gate. The use of a microscopic model, which takes account of the finite time and distance required for electrons to move into the upper valleys, correctly places the electron accumulation midway between the gate and drain. Hockney et al [52] and Moglestue [56] have also investigated the distrubution of electrons in the central and upper valleys for GaAs MESFETs, Figure 8.4. This has shown that the current in the channel below the gate of a 1 μm gate length FET is entirely due to central valley electrons, whereas in the gap between gate and drain the majority of carriers are low mobility upper valley electrons.

Time-dependent ac Monte Carlo simulations have also been performed on FET and diode models to investigate their dynamic behaviour under transient and microwave conditions. However, because of the excessive amounts of computer time required for a true a.c. analysis over several cycles, dynamic simulations are usually restricted to transient cases which make use of Fourier transform techniques to extract small-signal a.c. parameters [20] Despite the limitations in computer power, it is possible to conduct complete a.c. analyses for small simple device structures operating over several cycles at millimetric frequencies (around 100 *GHz*). An example of this type of simulation is given in the Schottky diode model of Beard and Rees [59].

More recent Monte Carlo simulations have been directed towards the study of heterojunction devices [58]. These models have included quantisation effects and have included self-consistent solutions of the Schrödinger's equations. These state-of-the-art simulations include detailed phonon and impurity scattering treatments, surface effects, electron-electron scattering and intersubband scattering. Degeneracy and impact ionisation are also included. Plasmon-electron and plasmon-coupled phonon scattering have been incorporated in some simulations. The results obtained from these simulations have been found to agree well with experimantal observations.

References

[1] Kurosawa, T., Proc. Int. Conf. on Physics of Semiconductors, Kyoto, (J.Phys. Soc. Japan, 21, Suppl. p.527), 1966

[2] Rees, H.D., "Calculation of steady state distribution function by exploiting stability", Phys. Lett., 26A, pp.416-417, 1968

[3] Rees, H.D., "Calculation of distribution function by exploiting the stability of the steady state", J. Phys. Chem. Solids, 30, pp.643-655, 1969

[4] Fawcett, W. and Rees, H.D., "Calculation of the hot electron diffusion rate for GaAs", Phys. Lett., 29A, pp.578-579, 1969

[5] Lebwohl, P.A. and Price, P.J., "Direct microscopic simulation of Gunn effect phenomena", Appl. Phys. Lett, 19, pp.530-532, 1971

[6] Price, P.J. and Stern, F., Surf. Phys., 132, pp.577-593, 1973

[7] Fawcett, W, "Electronic transport in crystalline solids", ed A. Salam, Vienna: International Atomic Energy Agency, pp.531-618, 1973

[8] Ruch, J.G., "Electron dynamics in short-channel field effect transistors", IEEE Trans. Electron Devices, ED-19, pp.652-659, 1972

[9] Baccarani, G., Jacoboni, C. and Mazzone, A.M., "Current transport in narrow-base transistors", Solid State Electron., 20, pp.5-10, 1977

[10] Hauser, J.R., Littlejohn, M.A. and Glisson, T.H., "Velocity-field relationship of InAs-InP alloys including the effects of alloy scattering", Appl. Phys. Lett., 28, pp.458-461, 1976

[11] Barker, J.R. and Ferry, D.K., "Self-scattering path-variable formulation of high-field, time-dependent, quantum kinetic equations for semiconductor transport in the finite-collision-duration regime", Phys. Rev. Lett., 42, pp.1779-1781, 1979

[12] Moglestue, C., "Monte Carlo particle simulation of hole-electron plasma formed in a pn junction", Electronics Letters, Vol.22, pp.397-398, 1986

[13] Pone, J-F., Castagne, R.C., Courat, J-P. and Arnodo, C., "Two-dimensional particle modeling of submicrometer gate GaAs FET's near pinchoff", IEEE Trans. Electron Devices, ED-29, pp.1244-1255, 1982

[14] Moglestue, C., "A Monte Carlo particle study of a semiconductor responding to a light pulse", Proc. Int. Conf. on Simulation of Semiconductor Devices and Processes, Swansea: Pineridge Press, pp.153-163, 1984

[15] Alberigi-Quarantra, A., Jacoboni, C. and Ottaviani, G., "Negative differential mobility in III-V and II-VI semiconducting compounds", Rev. Nuovo Cim., Vol.1, No.4, pp.445-495, 1971

[16] Price, P.J., Semiconductors and Semimetals, New York: Academic, 1979

[17] Jacoboni, C. and Reggiani, L., "Bulk hot-electron properties of cubic semiconductors", Adv. Phys., 28, pp.493-553, 1979

[18] Jacoboni, C. and Reggiani, L., "The Monte Carlo method for the solution of charge transport in semiconductors with applications to covalent materials", Rev. Mod. Phys., 55, pp.645-705, 1983

[19] Hockney, R.W. and Eastwood, J.W., Computer Simulation Using Particles, New York: McGraw-Hill, 1981

[20] Moglestue, C., "Monte Carlo particle modelling of small semiconductor devices", Comp. Meth. in Appl. Mech. and Engng., Vol. 30, pp.173-208, 1982

[21] Fawcett, W., Boardman, A.D. and Swain, S., "Monte Carlo determination of electron transport properties in gallium arsenide", J.Phys. Chem. Solids, 31, pp.1963-1990, 1970

[22] Birdsall, C.K. and Fuss, D., "Clouds in clouds, clouds in cells, physics for mnay-body plasma simulation", J. Comput. Phys., 3, p.494, 1969

[23] Eastwood, J.W. and Hockney, R.W., "Shaping the force law in 2D particle-mesh models", J. Comput. Phys., 16, p.342, 1974

[24] Hockney, R.W., Goel, S.P. and Eastwood, J.W., "Quite high resolution models of a plasma", J. Comput. Phys., 14, pp.148-158, 1974

[25] Hockney, R.W., "A fast direct solution of Poisson's equation using Fourier analysis", J.Assoc. Comput. Mach., 12, pp.95-113, 1965

[26] Ottaviani, G., Reggiani, L., Canali, C., Nava, F. and Alberigi-Quaranta, A., "Hole drift velocity in silicon", Phys. Rev. B, 12, pp.3318-3329, 1975

[27] Hockney, R.W., "Measurements of collision and heating times in a two-dimensional thermal computer plasma", J. Comput. Phys., 8, pp.19-44, 1971

[28] Brooks, H., "Theory of the electrical properties of germanium and silicon", Adv. Electron Phys., 7, pp.85-182, 1955

[29] Barker, J.R. and Ferry, D.K., "On the physics and modelling of small semiconductor devices - I, II and III", Solid State Electron., 23, pp.519-549, 1980

[30] Price, P.J., Proc. of 9th International Conference on the Physics of Semiconductors, ed S.M. Ryvkin, Leningrad: Nauka, p.753, 1968

[31] Price, P.J., "The theory of hot electrons", IBM J. Res. Dev., 14, p.12-24, 1970

[32] Aas, E.J. and Blotekjaer, K., "Monte-Carlo calculation of electron transport in polar semiconductors", J.Phys. Chem. Solids, 35, pp.1053-1059, 1974

[33] Lebwohl, P.A., "Monte Carlo simulation of response of a semiconductor to periodic perturbations", J. Appl. Phys., 44, pp.1744-1752, 1973

[34] Boardman, A.D., Fawcett, W. and Rees, H.D., "Monte Carlo calculation of the velocity-field relationship for gallium arsenide", Solid State Commun., 6, pp.305-307, 1968

[35] Ruch, J.G. and Fawcett, W., "Temperature dependence of the transport properties of gallium arsenide determined by the Monte Carlo method", J. Appl. Phys., 41, pp.3843-3849, 1970

[36] Maloney, T.J. and Frey, J., "Transient and steady-state electron transport properties og GaAs and InP", J. Appl. Phys., 48, pp.781-787, 1977

[37] Littlejohn, M.A., Hauser, J.R. and Glissan, T.H., "Velocity-field characterstics of GaAs with $\Gamma_6^c - L_6^c - X_6^c$ conduction band ordering", J. Appl. Phys., 48, pp.4587-4590, 1977

[38] Butcher, P.N., Fawcett, W. and Ogg, N., "Effect of field-dependent diffusion on stable domain propagation in the Gunn effect", Br. J. Appl. Phys., 18, p755-759, 1967

[39] Ohmi, T. and Hasuo, S., Proc. Int. Conf. on Physics and Semiconductors (USAEC), p.60, 1970

[40] Abe, M., Yanagisawa, S., Wada, O. and Takanashi, H., "Monte Carlo calculations of diffusion coefficient of hot electrons in n-type GaAs", Appl. Phys. Lett, 25, pp.674-675, 1974

[41] Bauhann, P.E., Haddad, G.L. and Masnari, N.A., "Comparison of the hot electron-diffusion rates for GaAs and InP", Electronics Letters, Vol.9, No.19, p.460-461, 1973

[42] Ferry, D.K., in Physics of Non-Linear Transport in Semiconductors, ed D.K. Ferry, J.R. Barker and C. Jacoboni, New York: Plenum, pp.577-588, 1980

[43] Fauquembergue, R.J., Zimmermann, A., Kaszynski, A., Constant, E. and Microondes, G., "Diffusion and the power spectral density and correlation function of velocity fluctuation for electrons in Si and GaAs by Monte Carlo method", J. Appl. Phys., 51, pp.1065-1071, 1980

[44] Jacoboni, C, "Generalisation of Fick's Law for non-local complex diffusion in semiconductors", Phys. Stat. Solidi B, 65, pp.61-65, 1974

[45] Jacoboni, C., Reggiani, L. and Brunetti, R., Proceedings of 3rd Int. Conf. on Hot Carriers in Semiconductors, J. Phys. (Paris) Colloquium, 42, c7-123, 1981

[46] Hill, G., Robson, P.N. and Fawcett, W., "Diffusion and the power spectral density of velocity fluctuations for electrons in InP by Monte Carlo methods", J. Appl. Phys., 50, pp.356-360, 1979

[47] Nougier, J.P., in Physics of Non-Linear Transport in Semiconductors, ed D.K. Ferry, J.R. Barker and C. Jacoboni, New York: Plenum, pp.415-465, 1980

[48] Reif, F, Fundamentals in Statistical and Thermal Physics, New York: McGraw Hill, 1965

[49] Lebwohl, P.A. and Price, P.J., "Hybrid method for hot electron calculations", Solid State Commun., 9, pp.1221-1224, 1971

[50] Curby, R.C. and Ferry, D.K., "Impact ionization in narrow gap semiconductors", Phys. Stat. Solidi A, pp.319-328, 1973

[51] Shichijo, H. and Hess, K., "Band-structure-dependent transport and impact ionisation in GaAs", Phys. Rev. B, 23, pp.4197-4207, 1981

[52] Hockney, R.W., Warriner, R.A. and Reiser, M, "Two-dimensional particle models in semiconductor analysis", Electronics Letters, 10, pp.484-486, 1974

[53] Park, Y-J, Tang, T-W and Navon, D-H, "Monte Carlo surface scattering simulation in MOSFET structures", IEEE Trans Electron Devices, ED-30, pp.1110-1115, 1983

[54] Warriner, R.A., "Computer simulation of gallium arsenide field-effect transistors using Monte Carlo methods", IEEE Solid St. Electron Devices, 1, pp.105-110, 1977

[55] Rees, H., Sanghera, G.S. and Warriner, R.A., "Low temperature F.E.T for low power high speed logic", Electron. Lett., 13, pp.156-158, 1977

[56] Moglestue, C., "Computer simulation of a dual gate GaAs field effect transistor using the Monte Carlo method", IEEE Solid St. Electron Devices, 3, pp.133-136, 1979

[57] Glisson, T.H., Hauser, J.R., Littlejohn, M.A., Hess, K., Streetman, B.G. and Shichijo, H., "Monte Carlo simulation of real-space electron transfer in GaAs-GaAlAs heterostructures", J. Appl. Phys., 51, pp.5445-5449, 1980

[58] Al-Mudares, M.A.R. and Ridley, B.K., "Monte Carlo simulation of scattering-induced negative differential resistance in AlGaAs/GaAs quantum wells", J. Phys. C: Solid State Phys., pp.3179-3192, 1986

[59] Beard, J.J. and Rees, H.D., "Hot electron effects in Schottky barrier mixers at high frequencies", Electron. Lett., 17, pp.811-812, 1981

CHAPTER 9

QUANTUM MECHANICAL EFFECTS
AN INTRODUCTION TO QUANTUM TRANSPORT THEORY

The validity of classical semiconductor simulation techniques is becoming increasingly questionable as device dimensions decrease towards a de Broglie wavelength (1000Å or less). New semiconductor modelling techniques which treat the wave nature of carriers are required which can account for quantum effects in these devices. The preceding chapters have dealt with semiconductor device models based almost entirely on the Boltzmann transport equation. This view of carrier transport is termed *semiclassical,* where the carrier motion during free flight due to an applied field is described by classical mechanics and scattering events are described by a quantum statistical treatment. *Classical transport physics,* is characterised by the description of carrier transport based on the use of a single carrier distribution function, which allows the macroscopic current flow to be calculated. There are many features of carrier transport which are neglected by the Boltzmann theory. Amongst the more important omissions are non-locality of scattering processes, strong scattering, dense systems, small systems, the effects of strong driving forces and the non-classical influence of driving fields. These phenomena are accounted for in quantum transport models. A complete quantum description requires a model based on *quantum transport theory.* Quantum transport theory has a completely independent origin and is both conceptually and mathematically more complex than the Boltzmann type of models [1,2,3,4,5]. It is important to distinguish between quantum transport theory which completely describes a quantum view of a system and the application of quantum mechanics to the semiclassical picture, where quantum mechanics are used to provide information on bulk and interface properties (for example [6,7]). Quantum transport theory is used to explain, support and set confidence limits for the

Boltzmann transport theory. It is also used to understand and develop genuine quantum transport phenomena.

Classical transport physics is based on the concept of a probability distribution function, which is defined over the phase space of the position and momentum of carriers and can be used to predict the macroscopic properties of the device (current flow etc). In contrast, as a consequence of Heisenberg's Uncertainty Principle, quantum transport theory does not allow the simultaneous specification of a carriers position and momentum. A quantum mechanical description of the system is given in terms of a probability density matrix, where the time evolution of the system is determined by Liouville-von Neumann equation.

The concept of the carrier distribution function is still meaningful for large devices (with critical dimensions greater than 1 μm), since the spatial variations in the distribution function occur over distances much larger than the de Broglie wavelength associated with the carriers. Also in addition to spatial considerations semiclassical transport models assume that collisons occur on a very short timescale compared with an observation time (relaxation times should be much shorter than transit times). In devices where the critical dimension is of the order of 0.25 μm the variations in the distribution function occur on a time scale comparable with the collision times, but still short compared with transit times. Hence, the distribution function is still useful in describing the behaviour of carriers, although the carrier transport does not stictly follow the Boltzmann transport theory. The concept of a classical distribution function is not valid for very small devices where the active channel is less than 250 $\AA$ long since the collision time is of the same order as the transit time. It is useful however to attempt to establish an equivalence to the distribution function.

A quantum mechanical analoguey to the classical concept of a phase-space distribution function is available in the form of the Wigner distribution [8]. An interesting feature of the Wigner function is that its associated equation of motion is very similar to the Boltzmann transport equation, within the constraints of the uncertainty principle. The Wigner function allows functions which describe statistical expectation values such as current density and carrier density to be extracted.

Although quantum transport models are continuing to be developed, a generalised computationally efficient description of quantum transport in devices is not yet available. Simplified treatments of quantum mechanics are available which provide a useful insight into device operation and are tractable using readily available computer resources [9]. It is not possible here to give a detailed description of quantum mechanical theory and it is recommended that if reader requires a greater insight he should consult one of the many texts now available on this subject for example Schiff, Dicke and Wittke, Bogolubov and Bogolubov. [10,11,12].

9.1. Extension of Semiclassical Transport Concepts to Quantum Structures

Device structures are already being developed which have dimensions comparable to the de Broglie wavelength (less than 1000 Å), which operate on a quantum interference basis. Present semiclassical models cannot adequately describe the operation of these devices, yet generalised full quantum transport models are not yet available. Future semiconductor modelling techniques, based on solutions to Schrödinger's equation, will treat electrons as waves propagating through the device. At the present time simplified models based on a treatment of quantum mechanical phenomena are beginning to emerge. In particular self-consistent solutions of the Poisson and Schrödinger equations using Monte Carlo methods and deterministic solutions, have been applied to a number of device structures.

9.1.1. Quantum Mechanics - Basic Concepts

In quantum mechanics a wave function $\psi(\mathbf{r},t)$, which is itself a function of position vector **r** and time, analogous to the classical trajectory $\mathbf{r}(t)$, is assumed to provide a complete quantum mechanical description of the behaviour of a particle with a given mass and potential energy. Qualitatively, the wave function is large where the particle is likely to be and small elsewhere. $\psi(\mathbf{r},t)$ is a solution to the three-dimensional single-particle time-dependent Schrödinger equation, which may be written in the form,

$$i\hbar\frac{\partial\psi}{\partial t} = -\frac{\hbar^2}{2m}\nabla^2\psi + V(\mathbf{r},t) \tag{9.1}$$

where $\hbar$ is the reduced Planck's Constant ($h/2\pi$), ψ is the wave function, m the particle mass, V potential energy and $\mathbf{r}$ is the position vector for the particle. The position probability density for the particle is

$$P(\mathbf{r},t) = \psi^*(\mathbf{r},t)\psi(\mathbf{r},t) = |\psi(\mathbf{r},t)|^2 \tag{9.2}$$

and

$$\int P(\mathbf{r},t)d^3r = 1 \tag{9.3}$$

which simply states that the probability of finding the particle somewhere in the region, defined by the three-dimensional volume element $dx\ dy\ dz$ (d^3r), is unity. The expectation value of the position vector $<\mathbf{r}>$ is defined as

$$<\mathbf{r}> = \int \mathbf{r}P(\mathbf{r},t)d^3r = \int \psi^*(\mathbf{r},t)\mathbf{r}\psi(\mathbf{r},t)d^3r \tag{9.4}$$

The expectation value of the position vector is the vector whose components are the weighted averages of the corresponding components of the position of the particle. The expectation values of other physical quantities, such as potential energy V may be obtained in a similar way if they are functions only of position vector $\mathbf{r}$.

The Schrödinger equation can be simplified by assuming that the potential energy $V(\mathbf{r})$ does not depend on time. If the wave function is written in the form

$$\psi(\mathbf{r},t) = u(\mathbf{r})f(t) \tag{9.5}$$

A general solution can be written as the sum of such separated solutions. Substituting into (9.1) and dividing through by the product,

$$i\frac{\hbar}{f}\frac{df}{dt} = \frac{1}{u}\left[-\frac{\hbar^2}{2m}\nabla^2 u + V(\mathbf{r})u\right] \tag{9.6}$$

The right-hand side of this equation depends only on $\mathbf{r}$ and the left-hand side only on t. Hence both sides must be equal to the same separation constant E. Integrating the equation for u becomes,

$$\left[-\frac{\hbar^2}{2m}\nabla^2 + V(\mathbf{r})\right]u(\mathbf{r}) = Eu(\mathbf{r}) \tag{9.7}$$

A solution of the wave equation may be obtained as,

$$\psi(\mathbf{r},t) = u(\mathbf{r})\exp\left(-i\frac{Et}{\hbar}\right) \tag{9.8}$$

The energy and momentum of a free particle can be represented by differential operators acting on the wave function ψ. Hence an expression of the total energy may be obtained from equation (9.9) as

$$i\hbar\frac{\partial\psi}{\partial t} = E\psi \tag{9.9}$$

This is an eigenvalue equation with ψ as the eigenfunction and E as the eigenvalue. The eigenvalue equation (9.7) has u and ψ as eigenfunctions of the operator $\left[-\frac{\hbar^2}{2m}\nabla^2 + V(\mathbf{r})\right]$, and an eigenvalue E.

It is often more convenient to represent the Schrödinger wave equation in matrix notation. Equation (9.7) may be re-written using hamiltonian notation as,

$$Hu_k(\mathbf{r}) = E_k u_k(\mathbf{r}) \tag{9.10}$$

where the subscript k denotes the different members of the orthonormal set of energy eigenfunctions $u_k(\mathbf{r})$ and the corresponding eigenvalues E_k. The Hamiltonian is given by,

$$H = \frac{\mathbf{p}^2}{2m} + V(\mathbf{r}) = -\frac{\hbar}{2m}\nabla^2 + V(\mathbf{r}) \tag{9.11}$$

It is convenient to write the Schrödinger equation in the form

$$i\hbar\frac{\partial\psi}{\partial t} = H_t\psi \tag{9.12}$$

The Hamiltonian may be divided into two parts

$$H = H_0 + H' \tag{9.13}$$

where H_0 does not depend explicitly on time and has a simple structure. H_0 is often chosen to represent the kinetic energy and H' the potential energy.

A particular state α of a system can be represented by a function $\psi_\alpha(\mathbf{r})$, which is a one column matrix where the rows are labelled by the co-ordinate $\mathbf{r}$. ψ_α may be regarded as a state vector in infinite-dimensional Hilbert Space [13]. Each dimension corresponds to one of the rows of the column matrix ψ_α. The component of the state vector along a particular axis of the Hilbert space is equal to the numerical value of the corresponding element of the matrix. State functions or state vectors are usually described using Dirac's bra and ket notation. The state function or state vector (for example ψ_α) is represented by a ket or ket vector $|\alpha>$ and the hermetian adjoint‡ state (for example $\psi_\alpha^\dagger$) by a bra or bra vector $<\alpha|$. The inner product of two state vectors, known as a bracket expression, is written as,

$$\psi_\alpha^\dagger \psi_\beta = <\alpha|\beta> \tag{9.14}$$

The time dependent Schrödinger equation may be written in terms of a time-dependent ket $|\alpha_s(t)>$ as,

$$i\hbar \frac{d}{dt}|\alpha_s(t)> = H|\alpha_s(t)> \tag{9.15}$$

The subscript s refers to the ket as viewed in the Schrödinger picture where the time variation obeys the ordinary differential equation of equation (9.15). The hermetian adjoint equation is given by

$$-i\hbar \frac{d}{dt}<\alpha_s(t)| = <\alpha_s(t)|H^\dagger = <\alpha_s(t)|H \tag{9.16}$$

since H is hermetian.

An arbitrary ket $|\alpha>$ can be written as a generalised sum of eigenstates of any operator Ω, for example

$$|\alpha> = \sum_\lambda{}' |\lambda><\lambda|\alpha> = \sum_\lambda{}' P_\lambda |\alpha> \tag{9.17}$$

where the summation $\sum_\lambda{}'$ implies a summation over a complete orthonormal set of states. The operator P_λ, which is defined as $|\lambda><\lambda|$, is known as a projection operator. P_λ projects out of $|\alpha>$ the part of the ket that is a particular eigenstate of Ω. Since Ω is hermetian, its eigenvalues are real and P_λ is also hermetian.

‡The hermetian adjoint $A^\dagger$ of a matrix A is the matrix obtained by interchanging rows and columns and taking the complex conjugate of each element. ie: if $B = A^\dagger$ then $B_{ij} = A_{ji}^*$. A matrix is hermetian (or self-adjoint) if it is equal to its hermetian adjoint. Hermitian matrices must be square matrices.

The equation of motion for the projection operator P_α, derived from equations (9.15) and (9.16) is given by,

$$i\hbar\frac{d}{dt}P_\alpha = i\hbar\left(\frac{d}{dt}|\alpha>\right)<\alpha| + i\hbar|\alpha>\left(\frac{d}{dt}<\alpha|\right)$$

$$= H|\alpha><\alpha| - |\alpha><\alpha|H = [H,P_\alpha] \qquad (9.18)$$

where the operator $[H,P_\alpha]$ is called a commutator bracket, defined as $[A,B] \equiv AB - BA$ (both matrices A and B must be square).

9.1.2. Application of Quantum Mechanics to Semiconductor Device Modelling

Monte Carlo models have been successfully applied to the study of AlGaAs/GaAs quantum wells used in current low-dimensional structures [14] by solving in a self-consistent manner the Poisson equation to obtain the potential distribution $V(x,y,z)$,

$$\nabla^2 V = -\frac{q}{\varepsilon_o\varepsilon_r}(N_D - n) \qquad (9.19)$$

and the time independent Schrödinger equation to obtain the wavefunction and subband energies,

$$\frac{\hbar^2}{2m}\nabla^2\psi + \psi_n(E_n + qV) = 0 \qquad (9.20)$$

where ψ_n is the wave function corresponding to subband n and E_n is the energy at the bottom of subband n. The electron distribution $n(z)$ is then obtained from

$$n(z) = \sum_n N_n |\psi_n(z)|^2 \qquad (9.21)$$

where N_n is the occupation of subband n by the given Fermi distribution. The process of calculating $V(x)$, ψ, E_n and n repeats until a converged self-consistent solution is obtained, where the potential obtained from the most recently obtained carrier density agrees, within the required limits of accuracy, with the previous potential substituted into the Schrödinger equation. The Monte Carlo simulation follows along the lines outlined in the previous chapter with the inclusion of the Schrödinger equation, using a many particle two-dimensional model. The wave-functions obtained from the solution of the Schrödinger equation are used to obtain the quasi-two-dimensional matrix elements for the scattering process. The free flight of electrons is modelled in the usual manner by applying the electric field and following the trajectories of the particles (each representing a cloud of electrons) subject to scattering events. Electron-electron scattering is treated separately [14,15,16]. A Monte Carlo quantum electrostatic simulation has been used by Al-Mudares and Ridley to study scattering induced negative differential resistance in GaAs/AlGaAs quantum wells, using planar two-dimensional models [14]. Results obtained from this simulation are shown in Figure 9.1.

A deterministic model for the self-consistent treatment of electron propagation in small scale semiconductor devices has been described by Cahay et al [17] who have their model on the original work of Tsu and Esaki [18,19]. This model neglects carrier scattering and assumes a simple band structure, which contrasts with the more detailed Monte Carlo model outlined above. The model obtains a self-consistent solution of the electron density and electrostatic potential by solving the wave equation in conjunction with the Poisson equation. A one-dimensional device is modelled, with two ohmic contacts in thermal equilibrium. The contacts launch electrons into the device with a range of wave-vectors **k**. The electron wavefunction in the device is given by

$$\Psi(\mathbf{r}) = \psi(x)\exp(ik_t . r_t) \tag{9.22}$$

which is obtained by solving

$$\frac{d}{dx}\left[\frac{1}{\gamma(x)} \cdot \frac{d\psi(x)}{dx}\right] =$$

$$\frac{2m^*(x_c)}{\hbar^2}[E_p + E_t(1 - E_t(1 - \gamma(x)^{-1}) - E_c(x)]\psi(x) = 0 \tag{9.23}$$

where $\psi(x)$ is the envelope function, x_c is the contact position, $\gamma(x)$ describes the spatial variation of the effective mass with respect to that in the contact $m^*(x_c)$ where

$$\gamma(x) = \frac{m^*(x)}{m^*(x_c)} \tag{9.24}$$

The transverse energy E_t and longitudinal energy E_p are given by

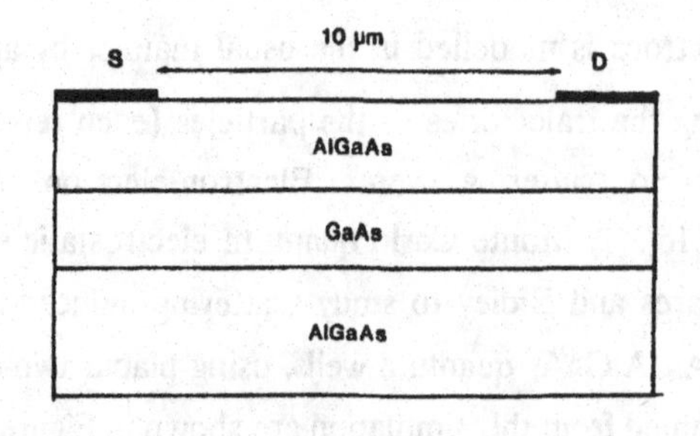

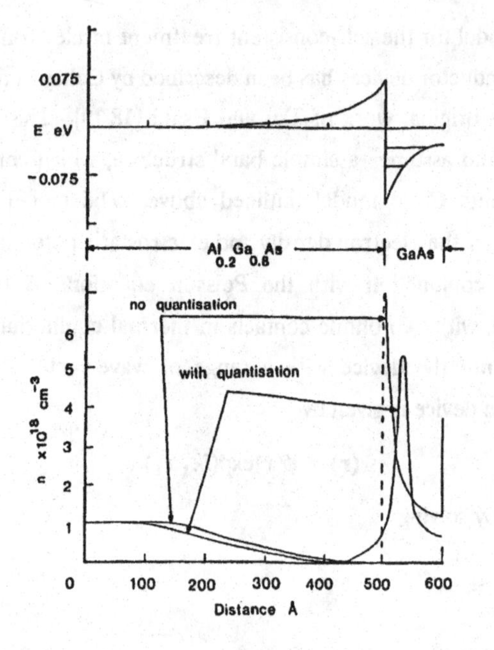

Figure 9.1 Results obtained from the simulation of GaAs/AlGaAs quantum wells [14]. (a) Double heterostructure used in the calculations (b) results showing band-bending and electron density distribution at an $Al_{0.2}Ga_{0.8}As-GaAs$ interface with $N_D = 10^{24}m^{-3}$.

$$E_t = \frac{\hbar^2 k_t^2}{2m^*(x_c)} \qquad E_p = \frac{\hbar^2 k_x^2}{2m^*(x_c)} \tag{9.25}$$

If it is assumed that the effective mass $m^*(x)$ is position independent, then $\psi(x)$

will be independent of $\mathbf{k}_t$ and the electron density for electrons impinging from the left will be described by,

$$n^{l-r}(x) = \frac{1}{2\pi}\int_0^\infty dk_x \ |\psi^+_{k_x}(x)|^2 \sigma(k_x) \qquad (9.26)$$

where $\sigma(k_x)$ is given by,

$$\sigma(k_x) = \frac{m^*(x_c)k_BT}{\pi\hbar^2} ln\left[1 + exp\left\{(E_{FL} - E_{CL} - E_p)/k_BT\right\}\right] \qquad (9.27)$$

where E_{FL} is the Fermi level at the left contact and k_BT is the thermal energy. The total electron density is obtained as a function of position by adding the contribution of the left contact to that due to the right contact. The integral in equation (9.26) is evaluated by summing the contributions to $n(x)$ for each k_x . This is achieved by incrementing the longitudinal wave-vector from zero to a maximum value and computing the wavefunction for each wavevector. The contribution for each k_x to the $n(x)$ for electrons between k_x and $k_x + dk_x$ is then evaluated. The corresponding integral for the other contact is treated in the same manner.

Equation (9.23) cannot be solved analytically for an arbitrary potential and is solved by discretising the potential over a finite number of steps between the contacts. Current continuity at each interface requires that $\psi(x)$ and $1/m^*(x)$ be continuous. $\psi(x)$ can be solved in a number of ways including the technique of cascading transfer matrices [18,19] and by using recursive methods [17,20].

The self-consitent solution for electron propagation is obtained by assuming an initial electrostatic potential $V(x)$ and then solving the wave equation to obtain $\psi(x)$ and $n(x)$. The Poisson equation is then solved again with the updated electron density n(x) to obtain a new value of $V(x)$. The new value of $V(x)$ is used to re-evaluate E_c which is then substituted into the wave equation and used to obtain an updated value of $n(x)$. The process is repeated until $V(x)$ calculated at successive steps converges to the required accuracy (usually less than twenty iterations).

Cahay et al have used this approach to model resonant tunneling structures consisting of AlGaAs/GaAs/AlGaAs layers [17]. The Poisson equation used in this example is of the form used to model heterojunctions described previously,

$$\frac{d}{dx}\left[\varepsilon_o\varepsilon_r(x)\frac{dV(x)}{dx}\right] = -q[N_D(x) - N_A(x) - n(x)] \tag{9.28}$$

where the relative permittivity ε_r is a function of material type and composition (a function of position x in this example). A simplistic treatment of the wave equation may assume that the conduction band profile is known. In the case of resonant tunelling structures, the electrons spend a great deal of time between confining barriers and the electron density can be large. In these circumstances the conduction band profile must be solved self-consistently using

$$E_C(x) = E_0 - \chi(x) - qV(x) \tag{9.29}$$

where E_0 is the field-free vacuum energy level and $\chi(x)$ is the electron affinity. Examples of results obtained for a resonant tunneling structure are shown in Figure 9.2.

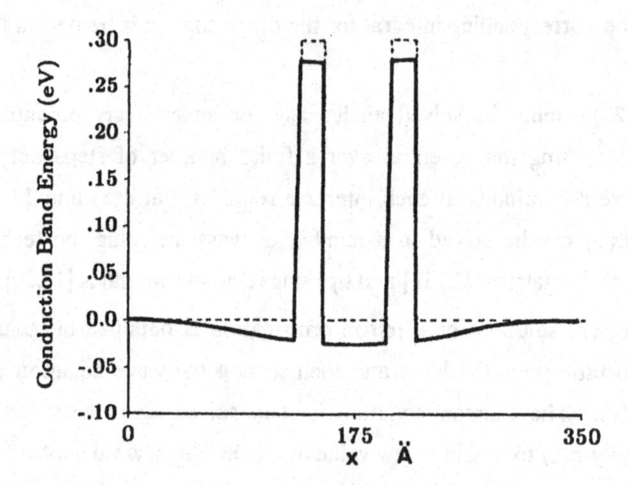

Figure 9.2 Results obtained from the simulation of a resonant tunnelling structure [17]. The calculated self-consistent energy band profile (solid line) for the resonant tunneling structure and assumed energy band profile (dashed line) are shown for comparison.

Detailed quantum mechanical models have been developed to study amorphous materials. In particular, amorphous silicon has been extensively studied [21,22,23,24]. Hickey et al [25] have obtained the density of states for amporphous silicon using the equation of motion method in **k** space. If the wave function is expanded as

$$\psi(\mathbf{r},t) = \Omega^{-1}\sum_{\mathbf{k}} a_{\mathbf{k}}(t)\exp(i\mathbf{k}.r) \tag{9.30}$$

where Ω is the volume of the system. The amplitudes satisfy the equation,

$$i\hbar\frac{\partial}{\partial t}a_{\mathbf{k}} - \frac{\hbar^2}{2m}k^2 a_{\mathbf{k}} - \sum_{\mathbf{k}} V(\mathbf{k}' - \mathbf{k})a_{\mathbf{k}'} = 0 \tag{9.31}$$

and

$$V(\mathbf{k}' - \mathbf{k}) = \Omega^{-1}\int d\mathbf{r}V(\mathbf{r})\exp(i(\mathbf{k}' - \mathbf{k}).r) \tag{9.32}$$

where $V(\mathbf{r})$ is the local potential. The model requires equation (9.30) to be solved by numerical integration subject to suitable initial conditions. The density of states is obtained by setting the amplitudes at $t = 0$ to have the form $\exp(i\phi_{\mathbf{k}})$ where $\phi_{\mathbf{k}}$ is a random phase. The course-grained density of states, as a function of energy E, is given by,

$$g(E) = \frac{1}{\pi\hbar}\mathrm{Re}\overline{\int_0^T \exp\left(-\frac{t}{\tau}\right)\exp\left(\frac{iEt}{\hbar}\right)\sum_{\mathbf{k}}\exp(-i\phi_{\mathbf{k}})a_{\mathbf{k}}(t)dt} \tag{9.33}$$

where $T \gg \tau$. The bar denotes averaging over all sets of random phases.

Studies of crystalline semiconductors have made extensive use of pseudopotentials to develop quantitative models. The localised bare pseudopotential of Guttman and Fong [23], has been used in several methods. Three pseudopotentials have been used in the work of Hickey et al [24]. These are the pseudopotentials ν_1 without exchange corrections, ν_2 which accounts for unscreened exchange and ν_3 which corresponds to the screened exchange approximation [26] The density of states using the pseudopotential ν_1 calculated for two structural models based on the Wooten, Winer and Waire structures [24], containing 216 atoms in a fully bonded network with periodic boundary conditions, are shown in Figure 9.3. The WWW1 structure was obtained using a Keating potential. The WWW2 structure introduces long range interactions by a generalisation of Weber's bond change model. The resolution of the calculations is 0.31 eV ($2\hbar/\tau$). Hickey et al have also calculated for the first time, the spectral function $\rho(kE)$ for amorphous silicon [24]. The spectral function is calculated in exactly the same way as the density of states, using equation (9.33), but setting only one coefficient a_k to be finite at $t=0$. The spectral function calculated using the pseudopotential ν_2 for the second Wooten, Winer and Waire structure is shown in Figure 9.4.

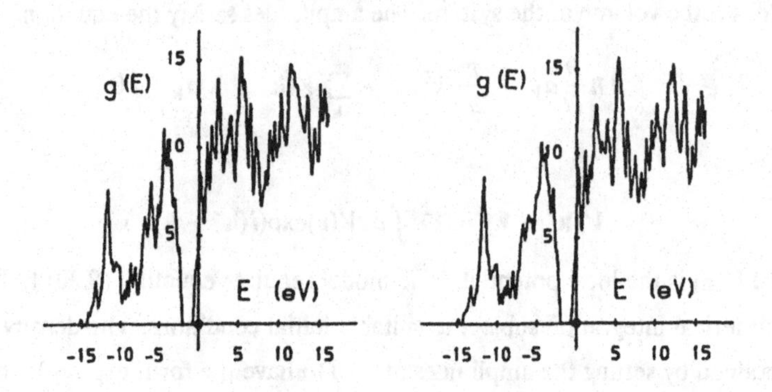

Figure 9.3 Density of states for a 216 atom fully bonded network [24]

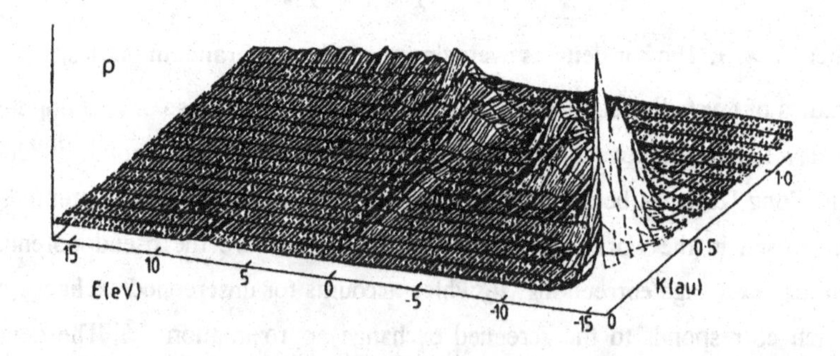

Figure 9.4 Spectral function calculated using the pseudopotential ν_2 for the WWW2 structure [24].

9.2. Quantum Transport Theory

The quantum mechanical theory discussed previously applies to systems in a pure quantum state, represented by a single ket $|\alpha>$. However, a complete knowledge of the state of the system is not usually available and it is necessary to adopt a statistical approach to describe the probability of the system being in particular states $|\alpha>, |\beta>$ etc. In classical transport physics, a pure classical state is represented by a point in phase space defined by its co-ordinates $r_1, r_2...r_n$ and momentum $p_1, p_2...p_n$ at every point in time. A statistical state is described by a non-negative density function $\rho(r_1, r_2,...r_n, p_1, p_2,...p_n, t)$ which determines the probability of finding the system in the interval $dr_1....dp_n$ at a particular time t. The quantum analogue of the classical density function is the density operator which is represented as a density matrix.

A statistical state, which describes a system which is not necessarily in a pure quantum state, is characterised by a set of probabilities p_α, p_β ... *etc* for being in the states $|\alpha>, |\beta>, ...$ *etc*, with random phase differences between their amplitudes. If one of the probabilities is equal to unity (and therefore all other probabilities are zero), the system is in a pure quantum state. If the states α are orthonormal (but not necessarily complete) then

$$\sum_\alpha p_\alpha = 1 \text{.....} \quad \textit{where } p_\alpha \geq 0 \tag{9.34}$$

The hermetian operator ρ is then defined as

$$\rho \equiv \sum_\alpha p_\alpha P_\alpha = \sum_\alpha |\alpha> p_\alpha < \alpha| \tag{9.35}$$

where P_α is the projection operator.

The rate of change of the probability density matrix ρ with time is determined by the Liouville-von Neumann equation, which in the Schrödinger picture is obtained from equation (9.18) and (9.35) as,

$$i\hbar \frac{d\rho}{dt} = [H, \rho] \tag{9.36}$$

This is the quantum analogue of the Liouville equation which governs the time evolution of the distribution function in classical transport physics where,

$$\frac{\partial \rho}{\partial t} = \{H,\rho\} \tag{9.37}$$

where the operator $\{H,\rho\}$ is known as a Poisson bracket†.

Quantum transport theory is based on the Liouville-von Neumann equation for the statistical density matrix ρ, which provides a complete description of the system

$$i\hbar \frac{\partial \rho}{\partial t} = \left[H_F,\rho\right] = H_F\rho - \rho H_F \tag{9.39}$$

where

$$H_F = H + F \tag{9.40}$$

where H is the Hamiltonian and F is the coupling to externally applied driving forces. For convenience the following equations are normalised with respect to $\hbar$. The usual boundary conditions for $t < 0$ is

$$\rho = \rho_o(H) \tag{9.41}$$

where ρ_o is a thermal equilibrium solution. The driving force is initiated at $t = 0$. The Hamiltonian H_F is usually partitioned into free carrier H_e, free scattering H_s, carrier-scatterer interaction H_{cs} and the electric and magnetic field driving force components (F) H_E and H_M respectively. Hence equation (9.40) expands to yield

$$H_F = H_e + H_s + H_{cs} + H_E + H_M = H_0 + H_{cs} + H_E + H_M \tag{9.42}$$

where

$$H_0 = H_e + H_s \tag{9.43}$$

Phenomenological effects such as current density J_i can be obtained by evaluating quantum-statistical expectation values determined from the density matrix $\rho(t)$. The expectation values of physical observables are obtained by taking the trace of the matrices, which are the products of the probability density matrix and the matrices that correspond to the physical observables. For example the current density is given as,

†The Poisson bracket {A,B} of any two functions of the co-ordinates q_i and momenta p_i is defined as

$$\{A,B\} = \sum_{i=1}^{f}\left(\frac{\partial A}{\partial q_i}\frac{\partial B}{\partial p_i} - \frac{\partial B}{\partial q_i}\frac{\partial A}{\partial p_i}\right) \tag{9.38}$$

$$<J_i(t)> = Tr\left[J_i\rho(t)\right] \tag{9.44}$$

and for a finite trace†.

$$<J_i(t)> \equiv \sum_{\lambda,\lambda'} <\lambda \,|\, J_i \,|\, \lambda'><\lambda' \,|\, \rho \,|\, \lambda> = \sum_{\lambda,\lambda'} J_i^{\lambda\lambda'} \rho_{\lambda'\lambda} \tag{9.46}$$

where $\{ \,|\, \lambda> \}$ is any complete set of states.

For a partitioned Hamiltonian H, the states $|\, \lambda>$ are chosen to diagonalize the term H_0, by defining,

$$|\, \lambda> \equiv \,|\, e> \,|\, s> \tag{9.47}$$

where the sets of states $\{|e>\}$ and $\{|s>\}$ diagonalise H_e and H_s respectively. It is important to note that the choice of $\{ \,|\, \lambda> \}$ determines the characteristics and interpretation of the subsequent transport theory.

The current density operator **J** may be expressed as

$$<\mathbf{J}(t)> = \sum J^{ee} f(e) \tag{9.48}$$

where it is assumed that the current-density depends only on electronic variables and commutes with the Hamiltonian component H_e. This is correct for homogeneous transport in the absence of magnetic fields. $f(e)$ is a real, time-dependent, generalised electron distribution function over the set of free-carrier states $\{ \,|\, e> \}$ of the form,

$$f(e) = <e \,|\, \sum_s <s \,|\, \rho \,|\, s> \,|\, e> \tag{9.49}$$

$$\equiv Tr_s <e \,|\, \rho \,|\, e> \equiv \ll e \,|\, \rho \,|\, e \gg_s \tag{9.50}$$

The Liouville equation (9.39) can be used in conjunction with $f(e)$ to produce a transport equation. The resulting equation may have a form similar to the Boltzmann equation, although the quantum nature of the states will affect the scattering rates. In contrast, for the cases of inhomogeneous transport and transport in quantizing magnetic fields, **J** is not diagonal and it is necessary to consider off-diagonal elements of the electron density matrix $f = <\rho>_s$. Although

†The trace of a matrix is the sum of the diagonal elements of the matrix,

$$Tr(A) = \sum_i A_{ii} \tag{9.45}$$

it is possible to express $f(e)$ in terms of $f(e,e')$, the resulting transport theory does not resemble the Boltzmann equation. It is found that in general a closed-form equation of motion for $f(t)$ can only be obtained for the case of independent carrier transport in a stationary scattering system. A generalised solution for the Louiville-von Neumann equation is not available. However, as only specific aspects of the full solution are of interest, it is possible to consider special cases such as single polaron (electron or hole) transport properties.

An alternative approach to interpreting quantum transport theory was proposed by Wigner [8] and has been extensively adopted (for example [27,28]). Wigner's distribution function has no simple interpretation in the sense of probability theory but can be used to directly calculate expectation values of observable quantities (current density etc). This is achieved in a manner analogous to classical transport theory by integrating the product of the Wigner distribution function and the observable over all phase space. For example, the current density $J(\mathbf{r},\mathbf{t})$ and carrier density $n(r,t)$ in a homogeneous system are given by the expressions

$$<\mathbf{J}(\mathbf{r},t)> = \sum_{\sigma} \int \frac{d^3\mathbf{k}}{(2\pi)^3} q\mathbf{v}(\mathbf{k}) f_\sigma(\mathbf{k},\mathbf{r},t) \tag{9.51}$$

$$<n(\mathbf{r},t)> = \sum_{\sigma} \int \frac{d^3\mathbf{k}}{(2\pi)^3} f_\sigma(\mathbf{k},\mathbf{r},t) \tag{9.52}$$

where $\mathbf{v}(\mathbf{k})$ is the group velocity of the electron momentum state $|\mathbf{k}>$ and $f_\sigma(\mathbf{k},\mathbf{r},t)$ is the Wigner one-electron distribution function defined as

$$f\sigma(\mathbf{k},\mathbf{r},t) \equiv \int d^3y \exp(-i\mathbf{k}.\mathbf{y})\, Tr\left\{\rho(t) V(x)^+_\sigma (\overline{r} - 1/2\mathbf{y})\, V(x)_\sigma (\mathbf{r} + 1/2\mathbf{y})\right\} \tag{9.53}$$

$$\equiv \int d^3\mathbf{y} \exp(-i\mathbf{k}.\mathbf{y})\, f_\sigma(\mathbf{r},\mathbf{y},t) \tag{9.54}$$

$\psi^+_\sigma(\mathbf{r})$ and $\psi_\sigma(\mathbf{r})$ are the second quantised creation and annihilation operators respectively, for a carrier of spin σ and position $\mathbf{r}$ [29]. ρ is second quantised and the trace is a many bodied trace. For homogeneous systems, where $f_\sigma(\mathbf{k},\mathbf{r})$ is independent of $\mathbf{r}$,

$$f(\mathbf{k},\mathbf{K},t) \equiv \int \exp(-i\mathbf{K}.\mathbf{R}) f_\sigma(\mathbf{k},\mathbf{R}) d^3\mathbf{R} \tag{9.55}$$

which is independent of K. In these circumstances the equivalent electron density matrix is diagonal in momentum space.

Wigner distributions do not always lend themselves to interpretation as probability densities since they are not necessarily positive definite. However, in the case of independent carriers f_σ reduces to the form $<r - 1/2y \mid <\rho>_s \mid r + 1/2y>$ where ρ is a fractional of the one-electron Hamiltonian [30].

The electron density matrix $f(t)$ is obtained from the full density matrix according to the relationship

$$f(t) = Tr_s \left[\rho\right] \tag{9.56}$$

For the case of stationary phonon and impurity distribution the initial thermal equilibrium density matrix may be expressed as

$$\rho_{t_0} \simeq f_o \left(H_e + V\right) \Omega_s \left(H_s\right) \tag{9.57}$$

Here f_o is of the Maxwellian equilibrium form and Ω_s describes the equilibrium distribution of scatterers. After some manipulation of the Liouville-von Neumann it is possible obtain an equation of the form [31].

$$\frac{\partial f}{\partial t} + i\left[H_e,f\right] + i\left[F,f\right] = i\int_0^t d\tau \hat{C}_f(\tau)f(t-\tau) + M_F(t) \tag{9.58}$$

The terms on the left-hand side of this equation describe collison-free diffusion and acceleration of carriers. The right-hand side is proportional to the scattering interaction. Collision and memory effects are accounted for by means of the terms C_f and $M_F(f)$ respectively. The memory term $M_F(t)$ results from the interaction between the electric field and scattering processes and is a function of the initial equilibrium state $f_o(H_e + V)$. In the homogeneous case f is a function of momentum only, and if

$$H_e = \frac{p^2}{2m^*} \tag{9.59}$$

then the term $\left[H_e,f\right]$ vanishes. In the context of this model [31], the coupling to the electric field **E**, obtained from the solution of Poisson's equation, is

$$F = -q\mathbf{E}.\mathbf{r} \tag{9.60}$$

and in the momentum representation i[F,f] reduces to $e\mathbf{E}\frac{\partial f}{\partial k}$. For the case of inhomogeneous transport the term $\left[H_e, f\right]$ gives rise to a component $\mathbf{v}(\mathbf{k})\frac{\partial f}{\partial \mathbf{r}}$.

It has been shown that equation (9.58) can be reduced to the Boltzmann transport equation for conditions of weak, infrequent scattering, translational invariance of the scattering system, point collisions and asymptotic time scale $t \gg \tau_c$ [32].

The influence of the electric field within a collision event causes the scattering integral to be modified in two important ways. The total energy-conserving δ function is broadened and the threshold energy of the emission of an optical phonon is modified causing a further shift in the energy δ function [31]. This feature of quantum transport theory is a major departure from the Boltzmann model and is known as the intra-collisional field effect (ICFE). The retardation effect of the ICFE on collisional interactions has been discussed by [33]. The collisional relaxation rate is retarded allowing a faster rise in the transient velocity response. The retardation in the collisional interaction also causes a faster rise in the electron temperature. This in turn results in a more rapid settling of the response to the steady state value. These effects have a significant influence on the operation of sub-micron semiconductor devices, increasing the frequency response and decreasing the distance over which the velocity transient occurs. In cases where quantum transport theory is used to investigate transport in semiconductors operating at very high frequencies, it becomes apparent that the collision-kernel is frequency dependent and that the collision response is largely controlled by the memory term M_F [30]. Boltzmann transport theory cannot account for these effects and is therefore invalid at high frequencies, even in homogeneous cases. Hence, it is important to account for the influence of quantum transport in devices with active channel lengths of less than 0.2 μm.

Generalised quantum transport theory which describes many-carrier, high field, inhomogeneous transport in a completely coupled hot carrier-hot-phonon-impurity system is still a relatively new area of research. Many-body quantum transport theory is conventionally analysed in the context of the Heisenberg uncertainty principle. The Wigner density matrices are evaluated as averages of appropriate annihilation and creation operators over the initial equilibrium density matrix $\rho(H)$. The dynamic evolution of the full system can be obtained from the time dependence of the field operators by coupling to the self-consistant local electric potential $V(\mathbf{r},t)$. Many-body quantum transport theory, includes Pauli degeneracy factors, generates coupled hot carrier and hot phonon transport equations, and has collision rates which are field dependent retarded functions of the Wigner electron distribution function and Wigner phonon distribution. It also introduces explicity and implicity field-dependent, self-consistent dynamic screening of the scattering processes which makes the electron kinetic equation non-linear in f.

A final aspect of quantum models which is of increasing importance when considering very small devices is the ability to analyse synergetic effects which arise from device-device interactions [34]. Complex integrated circuit systems with very small feature sizes (below 1000 $\AA$), may support synergetic processes. Many-body phase-transition theory provides a useful analogy with synergetic theory where multi-dimensional ($n > 1$) systems support ordered macrostructures with transitions between them. Synergetic effects in large systems such as VLSI structures could be investigated using a combination of available synergetic theory, nonlinear quantum transport theory and group theory [35]. Ferry and Barker have postulated that the quantum-mechanical operation of an array of very small devices may exhibit synergetic effects when the feature size is less than 0.1 μm [34].

9.2.1. Applications of Quantum Transport Theory

The inherent complexity of quantum transport theory and computational requirements for even relatively small systems, makes it difficult to apply directly to macroscopic device modelling in the way that the semiconductor equations and Monte Carlo methods are used (for example to extract time-domain terminal characteristics). However, it does provide an important means of characterising transport in very small semiconductor devices. Monte Carlo methods have been used to implement quantum transport theory for application to semiconductor device characterisation [36]. Semiconductor devices which have been investigated using quantum transport theory include short-channel MOSFETs and p-channel MESFETs [30]. Quantum transport theory has been used to demonstrate that the minimum practical limit for channels lengths in conventional MOSFETs is of the order of 0.24 μm [30].

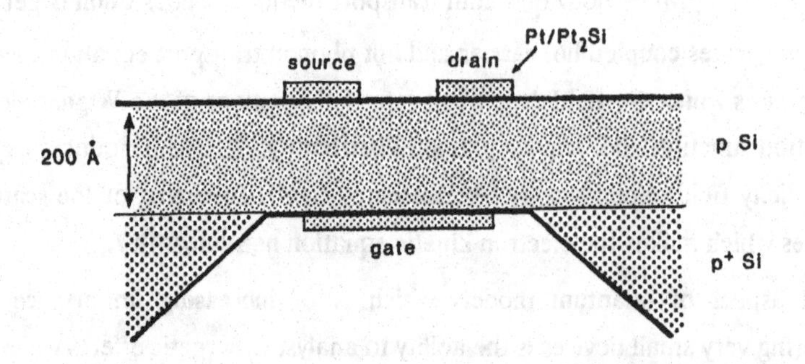

Figure 9.5 Small-scale p-channel MESFET structure [30]

A novel p-channel MESFET structure, shown in Figure 9.5, consisting of a thin 200 $\AA$, highly doped ($6 \times 10^{24} m^{-3}$) p type Si active channel with a Schottky barrier gate, has been described by Barker [30]. Ohmic contacts are not used for the source and drain in this device because quantum mechanical effects would otherwise lead to the channel becoming fully depleted. The current flow through the contacts is dominated by tunneling. This device has a predicted speed-power product of approximately $10^{-17} J$, which would make it an attractive proposition for future VLSI circuits.

Many-body theory has been used to confirm that the gallium arsenide non-equilibrium intervalley $(\Gamma - X)$ electron-phonon scattering rate de-screens in fields exceeding 1 kVm^{-1} [31].

Quantum transport theory has been used to characterise the Hall coefficient [37,38]. Morgan and Howson have developed linear response formulae for the Hall coefficient which depends on the states of the Fermi energy using a derivation based on the Wigner representation [39]. The analysis assumes that if the electric and magnetic field Hamiltonians H_E and H_M are switched on adiabatically so that they are proportional to $\exp(\varepsilon t)$, then the one-electron density matrix can be represented as,

$$\rho = f(H_0) + \rho_1 \exp(\varepsilon t) + \rho_2 \exp(2\varepsilon t) + ... \tag{9.61}$$

where f is the Fermi function, ρ_1 is proportional to the electric field E or the magnetic field H and ρ_2 is proportional to E^2, H^2 or EH. It is the latter contribution that is required to calculate the transverse electronic (Hall) conductivity σ_{xy} in the presence of a magnetic field. For the symmetric gauge, where the vector potential $\mathbf{A} = 1/2(H x \mathbf{r})$, and the magnetic field is in the z direction,

$$\sigma_{xy} = -\frac{2e}{\Omega}\sum_{ijk}(E_j - E_i - 2i\eta)^{-1}(E_k - E_i - i\eta)^{-1}V^*_{ij}H'_{jk}[H',f]_{ki}$$

$$- (E_j - E_k - i\eta)^{-1}V^*_{ij}[H',f]_{jk}H'_{ki} \tag{9.62}$$

where

$$\eta = \hbar\varepsilon \tag{9.63}$$

and

$$H' = H_E + H_m \tag{9.64}$$

E_i denotes and eigenvalue of H_0 and V_{ij} denotes a matrix element of the velocity operator $\mathbf{V}$. The electric field is applied in the y direction. The Hall mobility is obtained from equation (9.62) by considering those terms which are proportional to the product of H_E and H_M. Morgan and Howson have shown that σ_{xy} can be written as the sum of classical and non-classical components [39]. The classical component is a result of the magnetic field on the current generated by the electric field. The non-classical component is the result of the electric field acting on states created by the magnetic field. The latter component is zero for classical considerations.

The Hall coefficient R_H is given by,

$$R_H = \frac{\sigma_{xy}}{H\sigma_{xx}^2} \tag{9.65}$$

where H is the magnetic field in the z direction and the normal conductivity is given by Kubo's formula,

$$\sigma_{xx} = -\frac{2\hbar q^2}{\pi\Omega}\int dE\frac{\partial f}{\partial E}Tr(V_x G_I V_x G_I) \tag{9.66}$$

where

$$G_I = \pi\delta(E - H_0) \tag{9.67}$$

and V denotes the velocity operator here. Conductivities and Hall coefficients for amorphous and liquid transition metals, calculated using this quantum model, are shown as a function of Fermi energy in Figure 9.6.

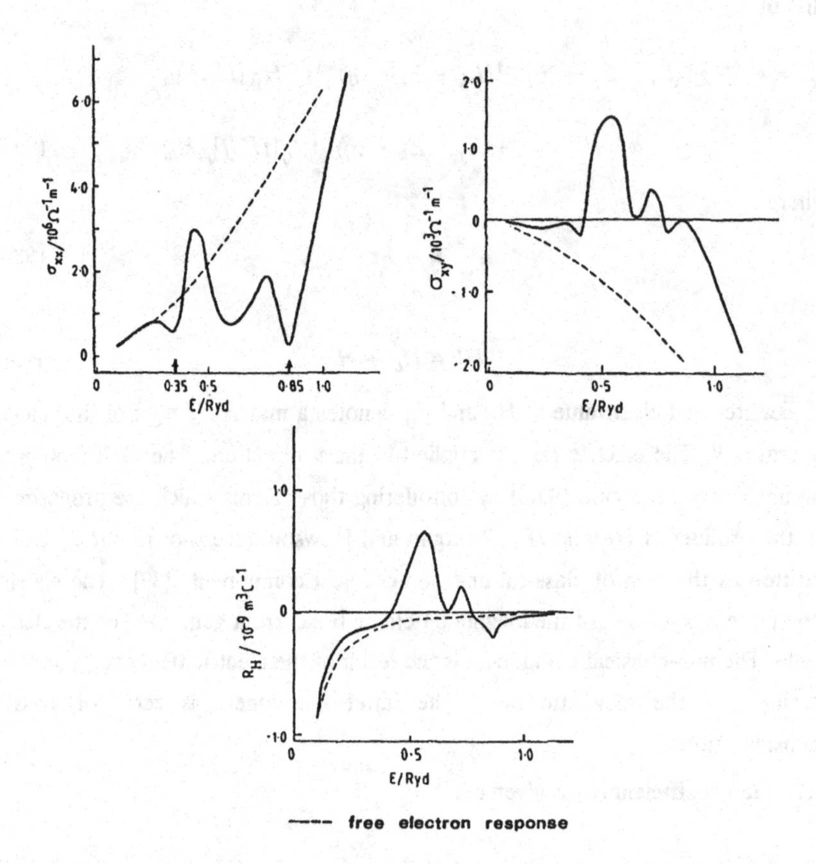

Figure 9.6 Calculated conductivities (σ_{xx} and σ_{xy}) and Hall coefficients (R_H) for amorphous and liquid transistion metals as a function of Fermi energy [39].

References

[1] Kubo, R, "Statistical-mechanical theory of irreversible processes I", J.Phys. Soc. Japan, 12, pp.570-586 ,1957

[2] Kohn, W. and Luttinger, J.M., "Quantum theory of electrical transport phenomena", Phys. Rev., 108, pp.590-611, 1957

[3] Luttinger, J.M. and Kohn, W., "Quantum theory of electrical transport phenomena II", Phys. Rev., 109, pp.1892-1909, 1958

[4] Chester, G.V., "The theory of irreversible processes", Rep. Prog. Phys., 26, pp.411-472, 1963

[5] Luttinger, J.M., Mathematical Models in Solid-State and Superfluid Physics, Edinburgh: Oliver and Boyd, 1968

[6] Price, P.J. and Stern, F., Surf. Phys., 132, pp.577-593, 1983

[7] Capassco, F., Surf. Sci., 132, pp.527-539, 1983

[8] Wigner, E., "On the quantum correction for thermodynamic equilibrium", Phys. Rev., 40, pp.749-759, 1932

[9] Gray, J.L. and Lundstrom, M.S., "Numerical solution of Poisson's equation with application in C-V analysis of heterojunction capacitors", IEEE Trans. Electron Devices, ED-32, pp.2101-2109, 1985

[10] Schiff, L.I., Quantum Mechanics, Tokyo: McGraw-Hill, International Edition, 1968

[11] Dicke, R.H. and Wittke, J.P., Introduction to Quantum Mechanics, Reading: Addison-Wesley, 1960

[12] Bogolubov, N.N. and Bogolubov, N.N. Jr., Introduction to Quantum Statistical Mechanics, Singapore: World Scientific Publ., 1982

[13] Halmos, P.R., Introduction to Hilbert Space, New York: Chelsea, 1957

[14] Al-Mudares, M.A.R. and Ridley, B.K., "Monte Carlo simulation of scattering-induced negative differential resistance in AlGaAs/GaAs quantum wells", J.Phys. C., 1986

[15] Takenaka, N., Inoue, M. and Inuishi, "Influence of inter-carrier scattering on hot electron distribution function in GaAs", J. Phys. Soc. Japan, 47, pp.861-868, 1979

[16] Inoue, M. and Frey, J., "Electron-electron interaction and screening effects in hot electron transport in GaAs", J.Appl. Phys., 51 , pp.4234-4239, 1980

[17] Cahay, M., Mclennan, M., Bandyopadhyay, S., Datta, S. and Lundstrom, M.S., "Self-consistent treatment of electron propagation in devices", Proc. 2nd Int. Conf. on Simulation of Semiconductor Devices and Processes, Swansea: Pineridge Press, pp.58-67, 1986

[18] Tsu, R. and Esaki, L., "Tunnelling in a finite superlattice", Appl. Phys. Lett., Vol.22, pp.562-564, 1973

[19] Chang, L.L., Esaki, L. and Tsu, R., "Resonant tunnelling of holes in AlAs-GaAs-AlAs heterostructures", Appl. Phys. Lett., Vol. 47, pp.415-417, 1985

[20] Koltun, M.M., Selective Optical Surfaces for Solar Energy Converters, New York: Allerton Press, 1981

[21] Ching, W.Y., Lin, C.C. and Guttman, L., "Structural disorder and electronic properties of amorphous silicon", Phys. Rev. B, 16, pp.5488-5498, 1977

[22] Bullett, D.W. and Kelly, M.J., "", J. Non-Cryst. Sol., 32, pp.225, 1979

[23] Guttman, L. and Fong, C.Y., "Self-consistent electronic structure of realistic models of amorphous hydrogenated silicon", Phys. Rev. B, 26, pp.6756-6775, 1982

[24] Hickey, B.J., Morgan, G.J., Weaire, D.L. and Wooten, F., "The density of states in amporphous Si", J. of Non-Cryst. Solids, 77, pp.67-70, 1985

[25] Hickey, B.J. and Morgan, G.J., "The density of states and spectral function in amorphous Si obtained using the equation of motion method k-space", to be published in J. of Non-Cryst. Solids, 1986

[26] Duggan, G., Morgan, G.J. and Lettington, A., "The effective screening function in Si, Ge and GaAs and the effects of exchange", Phys. Stat. Sol. B, 94, pp.659-665, 1979

[27] Tatarskii, V.I., "The Wigner representation of Quantum mechanics", Sov. Phys.-Usp., 26, pp.311-327, 1983

[28] Balazs, N.L. and Jennings, B.K., "Wigner's function and other distribution functions in mock phase spaces", Phys. Rep., 104, pp.348-391, 1984

[29] Hedin, L. and Lundquiest, S., Solid State Phys., 23, p1, 1969

[30] Barker, J.R., in Physics of Non-Linear Transport in Semiconductors, ed. K.Ferry, J.R.Barker and C.Jacoboni, New York: Plenum, pp.589-606, 1980

[31] Barker, J.R., in Physics of Non-Linear Transport in Semiconductors, ed. K.Ferry, J.R.Barker and C.Jacoboni, New York: Plenum, pp.127-151, 1980

[32] Barker, J.R., "Quantum transport theory of high-field conduction in semiconductors", J.Phys. C: Solid State Phys., 6, pp.2663-2684, 1973

[33] Ferry,in Physics of Non-Linear Transport in Semiconductors, ed. K.Ferry, J.R.Barker and C.Jacoboni, New York: Plenum, pp.577-588, 1980

[34] Barker, J.R. and Ferry, D.K, "On the physics and modelling of small semiconductor devices III - transient response in the finite collision-duration regime", Solid-State Electron., 23, pp.545-549, 1980

[35] Barker, J.R. and Ferry, J.K., Proc. Int. Conf. Cybernet. Soc., New York: IEEE Press, pp.762, 1979

[36] Barker, J.R. and Ferry, J.K, "Self-scattering path-variable formulation of high-field, time-dependent quantum kinetic equations for semiconductor transport in the finite-collision-duration regime", Phys. Rev. Lett., 42, pp.1779-1781, 1979

[37] Banyai, L. and Aldea, A., "Theory of the Hall effect in disordered system: impurity-band conduction", Phys. Rev., 143, pp.652-656, 1966

[38] Howson, M.A. and Morgan, G.J., "A model for the behaviour of the Hall coefficient in amorphous and liquid transition metals", Philosophical Magazine B, Vol. 51, No.4, pp.439-451, 1985

[39] Morgan, G.J. and Howson, M.A., "Linear response formulae for the Hall coefficient", J.Phys. C: Solid State Phys., 18, pp.4327-4334, 1985

APPENDIX 1
NUMERICAL SOLUTION OF THE CURRENT CONTINUITY EQUATION

The current continuity equation solutions are used to derive the electron and hole distributions n^{k+1} and p^{k+1} in terms of n^k, p^k, μ_n^k, μ_p^k, E^k, D_n^k and D_n^k, where the superscript k is the time step index. Two-dimensional current continuity algorithms based on linearised and Scharfetter-Gummel schemes are described below.

A.1 Linearised Continuity Scheme

The linearised semi-implicit time-dependent continuity scheme for electrons for a uniform mesh is,

$$\frac{n_{i,j,k+1} - n_{i,j,k}}{\Delta t} = \frac{1}{2q}\left[\nabla.J_{i,j}(n_{k+1},\mu_k E_k) + \nabla.J_{i,j}(n_k,\mu_k E_k)\right] \quad \text{(A.1.1)}$$

Δt is the time step and q the electronic charge. Expanding this expression,

$$n_{i,j,k+1,L+1} = \frac{\Delta t}{2q}\left[\frac{J_{x_{i+1/2,j}}(n_{k+1,L},\mu_k E_{x,k}) - J_{x_{i-1/2,j}}(n_{k+1,L},\mu_k E_{x,k})}{\Delta x}\right.$$

$$+ \frac{J_{y_{i,j+1/2}}(n_{k+1,L},\mu_k E_{y,k}) - J_{y_{i,j-1/2}}(n_{k+1,L},\mu_k E_{y,k})}{\Delta y}$$

$$+ \frac{J_{x_{i+1/2,j}}(n_k,\mu_k E_{x,k}) - J_{x_{i-1/2,j}}(n_k,\mu_k E_{x,k})}{\Delta x} \quad \text{(A.1.2)}$$

$$\left. + \frac{J_{y_{i,j+1/2}}(n_k,\mu_k E_{y,k}) - J_{y_{i,j-1/2}}(n_k,\mu_k E_{y,k})}{\Delta y}\right] + n_{i,j,k}$$

where the superscript L refers to the iteration index in calculating n^{k+1}, k is the time step index, i and j are the x and y spatial indices. Re-expressing this equation by substituting for J_x and J_y and incorporating under-relaxation, the implicit continuity scheme becomes,

$$n_{i,j,k+1,L+1} = \omega\left[\frac{\Delta t}{2}\left\{A_{i,j}n_{i+l,j,k+1,L} + B_{i,j}n_{i,j+1,k+1,L+1} + C_{i,j}n_{i-1,j,k+1,L+1}\right.\right.$$

$$\left.\left. + F_{i,j}n_{i,j-1,k+1,L+1} + G_{i,j}n_{i,j,k+1,L} + H_{i,j}\right\} + n_{i,j,k}\right]$$

$$- (\omega-1)n_{i,j,k+1,L} \quad (A.1.3)$$

where ω is the relaxation factor, which is less than unity for successive-under-relaxation (SUR). The factors in equation (A.1.3) are given by,

$$A_{i,j} = \frac{(a_{x_{i,j}} + b_{x_{i,j}})}{2\Delta x^2}$$

$$B_{i,j} = \frac{(a_{y_{i,j}} + b_{y_{i,j}})}{2\Delta y^2}$$

$$C_{i,j} = \frac{(b_{x_{i-1,j}} - a_{x_{i-1,j}})}{2\Delta x^2}$$

$$F_{i,j} = \frac{(b_{x_{i,j-1}} - a_{x_{i-1,j}})}{2\Delta x^2} \quad (A.1.4)$$

$$G_{i,j} = \frac{(a_{x_{i,j}} - a_{x_{i-1,j}} - b_{x_{i,j}} - b_{x_{i-1,j}})}{2\Delta x^2} + \frac{(a_{y_{i,j}} - a_{y_{i,j-1}} - b_{y_{i,j}} - b_{y_{i,j-1}})}{2\Delta y^2}$$

$$H_{i,j} = \frac{(c_{x_{i,j}} - c_{x_{i-1,j}})}{2\Delta x^2} + \frac{(c_{y_{i,j}} - c_{y_{i,j-1}})}{2\Delta y^2}$$

and

$$a_{x_{i,j}} = 1/2(\psi_{i,j} - \psi_{i+1,j})(\mu_{i,j} + \mu_{i+1,j})$$

$$a_{y_{i,j}} = 1/2(\psi_{i,j} - \psi_{i,j+1})(\mu_{i,j} + \mu_{i,j+1})$$

$$b_{x_{i,j}} = D_{i,j} + D_{i+1,j}$$

$$b_{y_{i,j}} = D_{i,j} + D_{i,j+1} \quad (A.1.5)$$

$$c_{x_{i,j}} = (n_{i,j,k} + n_{i+1,j,k}).a_{x_{i,j}} + b_{x_{i,j}}(n_{i+1,j,k} - n_{i,j,k})$$

$$c_{y_{i,j}} = (n_{i,j,k} + n_{i,j+1,k}).a_{y_{i,j}} + b_{y_{i,j}} (n_{i,j+1,k} - n_{i,j,k})$$

where n is the carrier concentration, D the diffusion coefficient, ψ the potential, μ the mobility, Δx the x space step and Δy the y space step.

This scheme is relatively easy to implement and gives satisfactory results in many applications. The accuracy of the above linearised scheme is discussed in Chapter 4, but is subject to significant errors for large time steps and in regions of rapid change in carrier concentration (or potential).

A.2 Steady-state 'Scharfetter-Gummel' Schemes

The popular Scharfetter-Gummel continuity scheme for the steady-state ($\partial n / \partial t = 0$) is expressed as:

For electrons:

$$\begin{aligned}
& n_{i,j-1}.D_{n_{i,j-1/2}}.B\left(\frac{\psi_{i,j-1} - \psi_{i,j}}{V_T}\right).\frac{a_{i-1} + a_i}{2b_{j-1}} \\
& + n_{i-1,j}.D_{n_{i-1/2,j}}.B\left(\frac{\psi_{i-1,j} - \psi_{i,j}}{V_T}\right).\frac{b_{j-1} + b_j}{2a_{i-1}} \\
& - n_{i,j}.\left[D_{n_{i,j-1/2}}.B\left(\frac{\psi_{i,j} - \psi_{i,j-1}}{V_T}\right).\frac{a_{i-1} + a_i}{2b_{j-1}}\right. \\
& + D_{n_{i-1/2,j}}.B\left(\frac{\psi_{i,j} - \psi_{i-1,j}}{V_T}\right).\frac{b_{j-1} + b_j}{2a_{i-1}} \\
& + D_{n_{i+1/2,j}}.B\left(\frac{\psi_{i,j} - \psi_{i+1,j}}{V_T}\right).\frac{b_{j-1} + b_j}{2a_i} \\
& \left. + D_{n_{i,j+1/2}}.B\left(\frac{\psi_{i,j} - \psi_{i,j+1}}{V_T}\right).\frac{a_{i-1} + a_i}{2b_j}\right] \\
& + n_{i+1,j}.D_{n_{i+1/2,j}}.B\left(\frac{\psi_{i+1,j} - \psi_{i,j}}{V_T}\right).\frac{b_{j-1} + b_j}{2a_i} \\
& + n_{i,j+1}.D_{n_{i,j+1/2}}.B\left(\frac{\psi_{i,j+1} - \psi_{i,j}}{V_T}\right).\frac{a_{i-1} + a_i}{2b_j} \\
& - G_{i,j}.\frac{a_{i-1} + a_i}{2}.\frac{b_{j-1} + b_j}{2} = 0 \qquad \text{(A.2.1)}
\end{aligned}$$

For Holes:

$$p_{i,j-1}.D_{p_{i,j-1/2}}.B\left(\frac{\psi_{i,j}-\psi_{i,j-1}}{V_T}\right).\frac{a_{i-1}+a_i}{2b_{j-1}}$$

$$+p_{i-1,j}.D_{p_{i-1/2,j}}.B\left(\frac{\psi_{i,j}-\psi_{i-1,j}}{V_T}\right).\frac{b_{j-1}+b_j}{2a_{i-1}}$$

$$-p_{i,j}.\left[D_{p_{i,j-1/2}}.B\left(\frac{\psi_{i,j-1}-\psi_{i,j}}{V_T}\right).\frac{a_{i-1}+a_i}{2b_{j-1}}\right.$$

$$+D_{p_{i-1/2,j}}.B\left(\frac{\psi_{i-1,j}-\psi_{i,j}}{V_T}\right).\frac{b_{j-1}+b_j}{2a_{i-1}}$$

$$+D_{p_{i+1/2,j}}.B\left(\frac{\psi_{i+1,j}-\psi_{i,j}}{V_T}\right).\frac{b_{j-1}+b_j}{2a_i}$$

$$+D_{p_{i,j+1/2}}.B\left(\frac{\psi_{i,j+1}-\psi_{i,j}}{V_T}\right).\frac{a_{i-1}+a_i}{2b_j}$$

$$+p_{i+1,j}.D_{p_{i+1/2,j}}.B\left(\frac{\psi_{i,j}-\psi_{i+1,j}}{V_T}\right).\left.\frac{b_{j-1}+b_j}{2a_i}\right]$$

$$+p_{i,j+1}.D_{p_{i,j+1/2}}.B\left(\frac{\psi_{i,j}-\psi_{i,j+1}}{V_T}\right).\frac{a_{i-1}+a_i}{2b_j}$$

$$-G_{i,j}.\frac{a_{i-1}+a_i}{2}.\frac{b_{j-1}+b_j}{2}=0 \qquad (A.2.2)$$

V_T is the scaled thermal voltage. The half-point values of the diffusion coefficients and carrier mobilities are obtained by linear interpolation.

$$D_{n_{i+1/2,j}}=\frac{D_{n_{i,j}}+D_{n_{i+1,j}}}{2} \qquad (A.2.3)$$

In order to avoid computational problems (underflows, overflows and inaccuracies), it is necessary to pay particular care in the implementation of the Bernoulli function. There are several possible approaches to obtaining values for the Bernoulli expressions. Simplified approximations are available, although if the normal exponential function available on the computer is chosen it is necessary to ensure that underflow and overflow errors do not occur. A useful method of implementing the Bernoulli function is to use the following scheme [A1]:

$$B(x) = \begin{cases} x \leq x_1 & -x \\ x_1 < x < x_2 & \dfrac{x}{exp(x)-1} \\ x_2 \leq x \leq x_3 & 1-\dfrac{x}{2} \\ x_3 < x < x_4 & \dfrac{x.exp(-x)}{1-exp(-x)} \\ x_4 \leq x < x_5 & x.exp(-x) \\ x_5 \leq x & 0 \end{cases} \tag{A.2.4}$$

The constants x_1 to x_5 depend on the individual computer hardware. They are defined using Selberherr's notation by:

$$\text{`}exp(x_1)\text{'}\ \text{`}-\text{'}\ 1\ \text{`}=\text{'}\ -1 \tag{A.2.5}$$

$$x_2\text{`}/\text{'}(\text{`}exp(x_2)\text{'}\ \text{`}-\text{'}1)\ \text{`}=\text{'}\ 1\ \text{`}-\text{'}\ (x_2\text{`}/\text{'}2)^{x_2} < 0 \tag{A.2.6}$$

$$1\ \text{`}-\text{'}\ (x_3\text{`}/\text{'}2)\ \text{`}=\text{'}\ x_3\ \text{`}.\text{'}\ exp(\text{`}-\text{'}x_3)\text{'}\ \text{`}/\text{'}(1\ \text{`}-\text{'}x_3)\text{'})^{x_3} > 0 \tag{A.2.7}$$

$$1\ \text{`}-\text{'}\ \text{`}exp(\text{`}-\text{'}x_4)\text{'}\ \text{`}=\text{'}\ 1 \tag{A.2.8}$$

$$\text{`}exp(\text{`}-\text{'}x_5)\text{'}\ \text{`}=\text{'}\ 0 \tag{A.2.9}$$

where the parameters inside quotation marks represent the computer implementation of the parameter.

A.3 Full Time-Dependent Schemes

The time dependent Scharfetter-Gummel continuity schemes are defined as:

For electrons:

$$n_{i,j-1,k+1}.D_{n_{i,j-1/2,k}}.B\left(\frac{\psi_{i,j-1,k+1}-\psi_{i,j,k+1}}{V_T}\right).\frac{a_{i-1}+a_i}{2b_{j-i}}$$

$$+n_{i-1,j,k+1}.D_{n_{i-1/2,j,k}}.B\left(\frac{\psi_{i-1,j,k+1}-\psi_{i,j,k+1}}{V_T}\right).\frac{b_{j-1}+b_j}{2a_{i-1}}$$

$$-n_{i,j,k+1}\cdot\left[D_{n_{i,j-1/2,k}}.B\left(\frac{\psi_{i,j,k+1}-\psi_{i,j-1,k+1}}{V_T}\right).\frac{a_{i-1}+a_i}{2b_{j-1}}\right.$$

$$+D_{n_{i-1/2,j,k}}.B\left(\frac{\psi_{i,j,k+1}-\psi_{i-1,j,k+1}}{V_T}\right).\frac{b_{j-1}+b_j}{2a_{i-1}}$$

$$+D_{n_{i+1/2,j,k}}.B\left(\frac{\psi_{i,j,k+1}-\psi_{i+1,j,k+1}}{V_T}\right).\frac{b_{j-1}+b_j}{2a_i}$$

$$+D_{n_{i,j+1/2,k}}.B\left(\frac{\psi_{i,j,k+1}-\psi_{i,j+1,k+1}}{V_T}\right).\frac{a_{i-1}+a_i}{2b_j}$$

$$\left.+\frac{1}{\Delta t}.\frac{a_{i-1}+a_i}{2}.\frac{b_{j-1}+b_j}{2}\right]$$

$$+n_{i+1,j,k+1}.D_{n_{i+1/2,j,k}}.B\left(\frac{\psi_{i+1,j,k+1}-\psi_{i,j,k+1}}{V_T}\right).\frac{b_{j-1}+b_j}{2a_i}$$

$$+n_{i,j+1,k+1}.D_{n_{i,j+1/2,k}}.B\left(\frac{\psi_{i,j+1,k+1}-\psi_{i,j,k+1}}{V_T}\right).\frac{a_{i-1}+a_i}{2b_j}$$

$$=\left(G_{i,j,k}-\frac{n_{i,j,k}}{\Delta t}\right).\frac{a_{i-1}+a_i}{2}.\frac{b_{j-1}+b_j}{2} \qquad \text{(A.3.1)}$$

For holes:

$$p_{i,j-1,k+1}.D_{p_{i,j-1/2,k}}.B\left(\frac{\psi_{i,j,k+1}-\psi_{i,j-1,k+1}}{V_T}\right).\frac{a_{i-1}+a_i}{2b_{j-1}}$$

$$+p_{i-1,j,k+1}.D_{p_{i-1/2,j,k}}.B\left(\frac{\psi_{i,j,k+1}-\psi_{i-1,j,k+1}}{V_T}\right).\frac{b_{j-1}+b_j}{2a_{i-1}}$$

$$-p_{i,j,k+1}.\left[D_{p_{i,j-1/2,k}}.B\left(\frac{\psi_{i,j-1,k+l}-\psi_{i,j,k+1}}{V_T}\right).\frac{a_{i-1}+a_i}{2b_{j-1}}\right.$$

$$+D_{p_{i-1/2,j,k}}.B\left(\frac{\psi_{i-1,j,k+1}-\psi_{i,j,k+1}}{V_T}\right).\frac{b_{j-1}+b_j}{2a_{i-1}}$$

$$+D_{p_{i+1/2,j,k}}.B\left(\frac{\psi_{i+1,j,k+1}-\psi_{i,j,k+1}}{V_T}\right).\frac{b_{j-1}+b_j}{2a_i}$$

$$+D_{p_{i,j+1/2,k}}.B\left(\frac{\psi_{i,j+1,k+1}-\psi_{i,j,k+1}}{V_T}\right).\frac{a_{i-1}+a_i}{2b_j}$$

$$+\frac{1}{\Delta t}.\frac{a_{i-1}+a_i}{2}.\frac{b_{j-1}+b_j}{2}$$

$$+p_{i+1,j,k+1} \cdot D_{p_{i+1/2,j,k}} \cdot B\left(\frac{\psi_{i,j,k+1} - \psi_{i+1,j,k+1}}{V_T}\right) \cdot \frac{b_{j-1} + b_j}{2a_i}$$

$$+p_{i,j+1,k+1} \cdot D_{p_{i,j+1/2,k}} \cdot B\left(\frac{\psi_{i,j,k+1} - \psi_{i,j+1,k+1}}{V_T}\right) \cdot \frac{a_{i-1} + a_i}{2b_j}$$

$$= \left(G_{i,j,k} - \frac{n_{i,j,k}}{\Delta t}\right) \cdot \frac{a_{i-1} + a_i}{2} \cdot \frac{b_{j-1} + b_j}{2} \qquad \text{(A.3.2)}$$

where k is the time step index.

References

[A1] Selberherr, S., Analysis and Simulation of Semiconductor Devices, Wien, New York:

INDEX

ADI method 73
affinity electron 138
aluminium gallium arsenide 135, 144
analysis 3, 37
anisotropy 66, 176
avalance breakdown 47

ballistic transport 129
band gap 34, 136, 150
band structure 136, 164
Bernoulli function 70, 152, 217
BJT bipolar junction transistor, 110
Boltzmann constant 16, 23
Boltzmann transport equation 14, 159
boundary condition, Dirichlet 28, 74
boundary condition, mixed 28, 75
boundary condition, Neumann 28, 75
boundary conditions 26, 74

carrier concentration 18
carrier heating 117, 127
carrier transport equation 14, 22
charge density 18
classical semiconductor equations 21, 24
collision term 16,17
conduction band 136, 162
conductivity, thermal 32
contact, ohmic 74
contact, Schottky barrier 89
continuity equation 18, 66, 106, 119, 214
Crank-Nicolson method 68
current density 18, 26
current density, electron, hole 18, 22, 122
cyclic Chebyshev method 72

Debye length 43
degeneracy 25
density of particles 14
density of states 25, 138
depletion capacitance 45
depletion region 40, 49, 90
diffusion 174
diffusion capacitance 46
diffusion coefficient 22, 120, 174
diffusion coefficient, thermal 33
diode, pn junction 38
discretization 64
displacement current 19
distribution function 14
    displaced Maxwellian 17, 161
divergence operator 18
drift velocity 23, 124, 173
drift-diffusion approximation 21, 131

Einstein relation 23
energy conservation 119
energy, electron 17, 120

Fermi energy 136, 138
Fermi function 197 ***
FET field effect transistor 47, 209
field electric 18, 19, 65
field magnetic 18
finite boxes 64
finite difference schemes 60
finite elements 100
flux, electric 19
flux, magnetic 18

Galerkin method 103
gallium arsenide 81, 88, 117, 126, 174, 177
Gaussian elimination 71
Gauss-Seidel method 71
generation 17, 29, 42
Gradual Channel Model 48
Green's theorem 105
Gummel algorithm 71

Hamiltonian 192, 202
HBT heterojunction bipolar transistor 149
heat, flow equation 33
heat, specific 33
Helmholtz equation 21
HEMT high electron mobility transistor 139
heterojunctions 135
Hilbert space 192
hydrodynamic equations 119

impact ionization 32, 178
indium phosphide 78, 118, 174, 177
interfaces 144

JFET junction field effect transistor 6,48
junction, abrupt 39, 45, 135
junction, linearly graded 40, 125

Laplace equation 54, 85
lifetime 17, 30
Liouville equation 16, 201
Liouville-von-Neumann equation 202
Lorentz force 16

mass, effective 127, 163
Maxwell's equations 17
MESFET 6, 88, 127, 130, 179
mesh 63, 90, 101, 108
mobility 21, 120, 151, 173
modelling, definition of 1
momentum conservation 16, 119
Monte Carlo method 155, 159, 194
MOSFET 6, 54, 82, 110

Newton polynomial 76
Newton-Raphson method 104, 107
normalised equations 24

packages, software 7, 85, 109
permittivity 19
phase space 14
pinch-off 51
Planck's constant 15
Poisson equation 21, 23, 25, 65, 105, 137, 162
potential, built-in 50, 89
potential, electrostatic 19
potential, quasi-Fermi 24
potential, vector magnetic 19
programs 7

quantum mechanical theory 188, 190
quantum transport theory 188, 201

recombination indirect 30
recombination 17, 29, 42
recombination, Auger 31
recombination, direct 30
recombination, optical 32
recombination, Skockley-Read-Hall 30
relaxation factor 72
relaxation time 17, 120, 126

scattering mechanisms 168
scattering probability 15, 165
Schrödinger equation 190, 194
screening 129
shape function 102
silicon 85, 110, 118, 128
simulation, definition of 1
SOR method 71, 73, 107
Stirling polynomial 75

Taylor series 61
TED transferred electron device 78
temperature, lattice 22
temperature electron 120, 123
thermionic emission 89
threshold voltage 55
timestep 69, 167
transconductance 51, 57
tunneling 47

valence band 136, 164
velocity electron 21, 123, 173
velocity group 15
velocity overshoot 117
velocity-field characteristics 174
VLSI 6, 112, 208

wave function 119
wave vector 15
Wigner distribution function 204

Zener breakdown 47